John Stewart

EVOLUTION'S ARROW

The Direction of Evolution and the Future of Humanity

The Chapman Press

First published in 2000 by The Chapman Press, PO Box 76 Rivett, Canberra, 2611 Australia.

National Library of Australia Cataloguing-in-publication data:

Stewart, John, 1952 Nov. 27-
Evolution's Arrow : the direction of evolution and the future of humanity.

Bibliography.
Includes index.
ISBN 0 646 39497 5.

1. Evolution. 2. Human evolution. I. Title.

599.938

For further articles and books by the author see:
http://users.tpg.com.au/users/jes999/

Contents

PART 1

Evolutionary Progress?

1

Introduction

The emergence of organisms who are conscious of the direction of evolution is one of the most important steps in the evolution of life on any planet. Once organisms discover the direction of evolution, they can use it to guide their own evolution. If they know where evolution is going, they can work out what will produce success in the future, and use this to plan how they will evolve.

Living things can evolve without having any knowledge of the direction of evolution. The diversity and complexity of life on earth is testimony to that. Organisms can try to deal with the future by blindly making changes to themselves or their offspring and seeing how the changes work out in practice. But this takes a lot of costly trial-and-error, particularly when the future is complex or changes rapidly. It is a bit like trying to drive a car through peak-hour traffic blindfolded. It will not be a winning strategy for organisms whose competitors can predict future events and use this to evolve more effectively.

The alternative is for organisms to guide their evolution by forming a picture of how evolution is likely to unfold in the future. They can try to find trends and patterns in this evolution that might impact on their future chances of survival. They can then use these patterns to see how they must change themselves and the way they are organised in order to continue to be successful.

On this planet, the organism that appears likely to take this significant evolutionary step is us. Our growing understanding of evolution is providing us with the knowledge that will enable us to see that there are large-scale patterns in the evolution of life. And it is a short step from this to recognising the evolutionary significance of using these patterns to guide our own evolution. But this significant step will not be possible until we have developed a comprehensive understanding of the direction of evolution and of its implications for humanity. The development of this theory will itself be an important step in our evolution. Key issues that it will have to address include:

- what is the direction of evolution? Where is it headed? Is the direction of change progressive, in the sense that life advances and improves as evolution unfolds? If it does progress, in what way do organisms improve?
- where does humanity fit in? Are we to be like the dinosaurs, better than what has gone before us, but soon to be replaced by something superior? Or can we play a significant role in the future evolution of life in the universe?
- what choices do we have? If we can see where evolution is going, is it possible for us to change to fit in with the direction, so that we can survive to participate in the next steps in evolution? Or should we ignore the direction of evolution, and live our lives in ways that might make us irrelevant to future evolution? Can we turn our back on the evolutionary processes that have produced us?
- what does this mean for us as individuals, here and now, for the way we live our lives and the way we organise ourselves socially? If we decide to do what we can to ensure that humanity participates in the future progressive evolution of life in the universe, what do we have to do, individually and collectively? Will we have to change our economic and social systems? Our psychology?
- Can deeper understandings of evolution and of its direction assist in answering the ancient questions that confront all aware human beings: where do we come from? what are we? where are we going to? Is there a purpose to human existence?

These are the central themes of this book.

In the chapters that follow, I will argue that evolution has direction, and that the direction is progressive. I will also show that this direction is important in answering the fundamental question of how we should live our lives. Awareness of the direction of evolution is capable of providing direction to our lives and for humanity as a whole.

To clear up one point of possible confusion immediately, I will be showing that evolutionary change progresses in evolutionary terms, not in human terms. Organisms improve as evolution unfolds in the sense that they become more competitive and better adapted than those they replace. They get better at surviving. But they will not necessarily get better against criteria that are important to humans. For example, the competitiveness of an animal might increase if it develops the ability to physically terrorise other members of its species to get a greater share of food. This would be an improvement in evolutionary terms, but many of us would not consider it to be progress in human terms[1].

This distinction is particularly clear in human evolution. The idea that human society progresses has taken a battering in the 20th century. We have seen the largest scale wars in human history. Six million Jews and twenty million Russians died in the Second World War alone. And modern societies have not necessarily produced better lives for their citizens. Members of earlier tribal societies arguably experienced happier and more meaningful lives than members of technologically advanced nations.

Much change in human society has not been progressive in human terms. But this does not mean humanity has failed to progress in evolutionary terms. Most of us now live in nation states that have proved their evolutionary superiority to tribal societies by replacing them over most of the planet.

There is a further important difference between these two different types of progress. The criteria used to assess whether evolutionary progress has occurred in any instance are objective. If organisms have improved their competitiveness and their adaptive fit to their environment, they have progressed in evolutionary terms. This may be difficult to assess in practice. But it is not fundamentally subjective like deciding the criteria that should be used to assess whether change is progressive in human terms. There are as many ways of defining progress in human terms as there are different sets of human values.

Our ability to assess objectively whether evolution progresses does not mean the issue is free of controversy. Evolutionists do not currently agree on whether evolution is progressive. Most believe it is not. The view that evolution is progressive and that humans are now at the leading edge of evolution on this planet is not supported by most evolutionary thinkers[2]. A major task of this book will be to show that they are wrong.

Progressionist ideas about evolution were popular until the middle of this century[3], but have since come under increasing attack. This is largely because progressionists have been unable to identify any plausible evolutionary mechanism that would continually drive progressive change along some absolute scale.

Anti-progressionists such as the noted American evolutionary writer Stephen Jay Gould argue that there is no such mechanism. They say that current evolutionary theory does not include any process that would produce general and on-going improvement as life evolves[4]. Natural selection adapts populations of organisms only to the specific local circumstances encountered by each population. This may produce some short-term improvement and directional change as the organism adapts better to local conditions, or as the environment changes. For example, a population of snow hares might progressively evolve thicker fur if average winter temperatures increase from year to year. But the directional change will end when the opportunity for improvements in local adaptations is exhausted, or when the local environment changes again in some other direction. And, Gould argues, better adaptation to local conditions will not produce general advance or progress. Changes that adapt a particular organism to its specific environment would not improve it for many other environments. A fish has no use for a better wing, or a bird for more efficient gills. Gould cannot envisage improvements that would be better in all conditions.

Gould and his supporters conclude that the earlier enthusiasm for progressionist views has no sound evolutionary basis: there is no mechanism within evolution that drives on-going progress; natural selection is a process that produces only local adaptation, not general advance or progress; and both the fossil record and the pattern of life we see about us are consistent with this. According to Gould, the belief that humans are at the leading edge of evolution is best explained as wishful thinking.

Progressionist views are currently in a similar position to evolutionary ideas prior to Darwin's *The Origin of Species*. In the centuries before Darwin, many thinkers had come up with the idea that some form of evolution best explained the pattern of plants, animals and fossils they saw in nature[5]. But they could not identify a plausible mechanism that explained how this evolution could occur. Evolutionists prior to Darwin could demonstrate that some aspects of the pattern of nature were consistent with evolution, but this consistency could easily be dismissed as lacking any causal basis. Like patterns of stars in the night sky that resemble shapes significant to humans, the consistencies could be dismissed as the product of creative imagination, not the result of real, causal relationships.

Darwin's great contribution was not the idea of evolution. It was to identify natural selection as the cause of evolution, and to demonstrate that natural selection was the inevitable result of sequences of tangible, concrete events in nature. He showed that evolution would occur wherever some organisms were more successful than others due to differences that could be passed to their offspring.

If progressionist views are to gain the widespread acceptance achieved by evolutionary theory, progressionists have to meet the challenge of identifying a concrete, causal basis for evolutionary progress. Without this, it is not possible to distinguish between patterns and trends that are accidental and meaningless, and those that are necessary and real. A central task of this book is to meet this challenge by showing that evolution includes processes that drive it in a particular direction, and that the direction is progressive.

I will show that the direction of evolution is towards increasing cooperation between living organisms. As evolution proceeds, living things will increasingly coordinate their actions for the benefit of the group rather than acting only in their own individual interests. Cooperators will inherit the earth, and eventually the universe[6].

Part Two of this book (Chapters 3 to 7 inclusive) is devoted to demonstrating that evolution is progressive, and that it produces increasing cooperation amongst living processes. Part of the argument in favour of this position is not controversial: it is beyond doubt that cooperation can be efficient and effective in evolutionary terms. Whatever challenges organisms face during their evolution, the challenges can be met more effectively through cooperation.

But how can evolution progress by exploiting these benefits of cooperation when, as Richard Dawkins and others have shown so clearly[7], evolution favours organisms that put their own selfish interests above all else? We will see that there is a solution to this apparent paradox: cooperation can flourish without organisms giving up their self-interest. Organisms can be organised so that beneficial cooperation is also consistent with their self-interest. When organisms are organised in this way, it is in their interests to be cooperative.

Within such an organisation, individuals will benefit when they cooperate, and will be harmed if they hurt the organisation. An example is a human business that is organised so that employees who work together to develop a new product obtain a share in any profit or loss it produces. In a group organised in this way, individuals who follow their own interests will also generally serve the interests of the group. Wherever cooperation pays off for the group, cooperation will generally also be in the interests of its members.

Evolution progresses towards greater cooperation by discovering ways to build cooperative organisations out of components that are self-interested. It has done so repeatedly throughout the history of life on earth. Cooperative groups of self-replicating molecular processes formed the first simple cells, groups of these cells formed larger and more complex cells, these in turn formed cooperative groups of cells that became multicellular organisms, and groups of multicellular organisms formed cooperative insect societies and human social systems.

One thing that is striking about this is that the cooperative groups that arose at each step in the sequence became the organisms that then teamed up to form the cooperative groups (and organisms) at the next step in the sequence. The result has been that all larger-scale living organisms are made up of smaller-scale living processes that are in turn made up of still smaller-scale processes and so on. And for the organism to operate effectively, all these layers of living processes must cooperate in the interests of the organism. All organisms, each of us included, are cooperative organisations.

It is also obvious that this sequence has direction. As the sequence has unfolded, the scale over which living processes cooperate has increased. In the evolution of life on this planet, cooperation between living processes began over very small scales and has progressively increased through the formation of larger and larger-scale cooperative groups and organisms. And in the last 10,000 years, this trend has accelerated enormously. Cooperative human groups have increased in scale from small tribal societies to nation states and empires, and now to forms of human organisation that operate on the scale of the planet (e.g. multinational companies, and economic markets).

Three thousand million years ago, cooperation extended only between molecular processes that were separated by about a millionth of a metre, the scale of early cells. Now, cooperation extends between human organisms that are separated by up to 12 million metres, the scale of the planet. And by up to 380 million metres when there are moon landings. Cooperation between molecular processes and cells now also extends over these larger scales. When humans cooperate in world-wide economic activities, so do their cells. And these increases in the scale of cooperation are unlikely to end here and now. The same evolutionary forces that drove the expansion of cooperative organisation in the past can be expected to continue to do so.

We will see that evolution progresses towards increasing cooperation whatever the mechanism that produces evolutionary change in organisms. As long as the mechanism is good enough at finding better adaptation, it will discover and exploit the benefits of cooperation. Both natural selection and the processes that produce cultural evolution in humans produce progressive evolution. Furthermore, the advantages of cooperation can be expected to drive progressive evolution wherever life emerges. On any planet where life evolves, evolution can be expected to produce cooperation over wider and wider scales, as it has on earth.

To exploit the benefits of cooperation effectively, groups of entities must evolve an ability to discover the most useful forms of cooperation, and to modify them as conditions change. They must be able to evolve and adapt their cooperation. The better and quicker they are at discovering effective cooperation, the better they will do in evolutionary terms. Imagine the evolutionary success enjoyed by the first groups of molecular processes to discover how to cooperate to construct a cellular membrane, the first groups of cells to produce a network of nerves to coordinate their activities, and the first groups of humans to learn how to chase game animals off cliffs.

So it is not only through increases in cooperation that evolution progresses. It also progresses through increases in the ability of living processes to adapt and evolve. The advantages of being better at adapting have driven progressive improvements in the evolvability and adaptability of cooperative groups, and of the organisms they eventually produce. The processes that adapt and evolve organisms have got progressively better at discovering the most effective forms of cooperation amongst the living processes that make up the organisms. Evolution itself evolves, and living processes get smarter at evolving.

Part Three of the book (Chapters 8 to 12 inclusive) deals with this evolution of the processes that adapt and evolve living processes. We will see how progressive evolution has improved the ability of the genetic evolutionary mechanism itself to adapt organisms[8]. The genetic mechanism uses trial-and-error to search for better adaptation. It tries out genetic changes when offspring are produced. If a change improves the ability of an offspring to survive and reproduce, it spreads throughout the population, producing organisms that are better adapted.

If these genetic changes are made randomly, the majority will be harmful. Most changes made blindly to a complex organism will kill it. Random change is a very inefficient way to search for improvements. So a genetic mechanism that can target its changes will have an evolutionary advantage. It can cut down on the number of changes that are harmful, and make changes that have a greater chance of being useful. For example, a population of snow hares in an environment in which general temperatures are changing widely every few years could target its genetic changes at varying the thickness of fur. This would be more likely to pay off than changing genes that are unrelated to environmental changes. A genetic mechanism that can focus genetic change in this way would be more efficient at discovering better adaptation. We will see that important features of genetic systems have evolved to target genetic change. We will see that sexual reproduction itself owes its existence to its ability to do this. Sex is smart[9].

But improvements in the genetic evolutionary mechanism can go only so far in enhancing the ability of organisms to adapt and evolve. The genetic mechanism can try out changes and discover better adaptation only when organisms reproduce. It cannot search for improvements during the life of the organism. Entirely new adaptive and evolutionary mechanisms had to be developed to exploit the great advantages of doing this. The new mechanisms had to be able to try out and test changes within the organism during its life[10]. Cells, multicellular organisms and human societies have all evolved internal processes that discover new and better adaptation in this way. Typical examples are our own physiological, emotional and mental adaptive systems.

We will see how the advantages of smarter adaptability and evolvability have driven a long sequence of improvements in these internal adaptive processes. A key milestone was reached when organisms could communicate with each other about adaptive improvements they had discovered during their lives. Adaptive discoveries no longer died with the individual who made them. They could be passed on to others, and a culture of adaptive information could be developed. Once this ability evolved, the internal adaptive processes qualified as evolutionary mechanisms, able to accumulate adaptive discoveries across the generations. On this planet, only humans and our societies have evolved this capacity to a high level.

A further key milestone in the progressive improvement of evolvability was the development of a capacity for mental modelling. Again, on this planet only humans have fully developed this ability. An organism capable of mental modelling can form internal mental models and mental pictures of how its environment will unfold in the future, and how its actions will affect this. To an extent, it can predict the future. So it is able to try out possible actions mentally, select the one that produces the best future result in its mental models, and only then try it out in practice[11]. It will be able to use its mental models to see how to manipulate its environment to achieve its particular objectives.

Over the generations, organisms with this capacity can collect more and more knowledge about their environment and the effects of their actions. This will enable them to build mental models of their environment that are more comprehensive and accurate. Progressively the organisms will be able to model how their environment unfolds over wider and wider scales of space and time. Eventually the organisms will be able to model the wider-scale evolutionary processes that have produced them and that will affect them in the future. For the first time they will see themselves as situated at a particular point in an on-going and progressive evolutionary process. And they will not just become aware of the direction of evolution. They will also become aware that their increasing awareness of the direction of evolution is itself a significant step in evolution.

The organisms will see that their existing physical adaptations and their existing motivations, interests, beliefs, and values are all the products of their evolutionary history. These characteristics will have all been tailored and tuned by past evolution to ensure that the organism survives. As their understanding of the direction of evolution improves, they will also see what they will have to do in the future to continue their evolutionary success. The organisms will see what they have to do both as individuals and socially: they will understand that they must further exploit the benefits of cooperation by forming cooperative organisations of larger and larger scale and greater and greater evolvability.

But will the organisms use their awareness of the direction of evolution to guide their own evolution? Will they choose to do what is necessary for future evolutionary success? Will they care about their evolutionary future? The difficulty faced by all organisms at this stage in their evolution is that they will be unlikely to find satisfaction and motivation in what they have to do for future evolutionary success. Continued success will demand radical changes in their behaviour and social organisation. But their existing motivations, moral codes, and values will influence their willingness to make these changes. The problem is that their motivations and other predispositions will have been moulded by the needs of past evolution, not future evolution.

Past evolution will have tailored their motivations and values so that they find satisfaction in behaviours and actions that would have produced success in the past, not those that will produce success in the future. Up until the development of their capacity for mental modelling, they will have been adapted by evolutionary mechanisms that were without foresight, and could not take into account the needs of future evolution.

It is one thing for an organism to know what it has to do for future evolutionary success. It is another thing entirely to want to do something about it. It is a bit like a person who knows that it is in his longer-term interests to work long hours and save money while he is young to provide for a comfortable retirement. As many of us know, awareness of our longer-term interests will not automatically motivate us to do what is necessary to serve those interests. The difficulty in finding motivation to pursue future evolutionary success is even greater. The individual will often not benefit at all by pursuing evolutionary objectives. It often will be only future generations who do so.

A better understanding of this difficulty can be had by imagining the following scenario: you are able to travel back in time, and you have been given the job of going back 50,000 years to show a band of human hunter-gatherers how they must change to achieve future evolutionary success. You are to use your knowledge of how evolution has unfolded since then to get them to change in the ways necessary for them to achieve future success. How likely are you to get them to change? Would they freely choose to reorganise themselves in the ways that have proven successful for human groups since then? For example, would they want to band together with other tribes, give up their nomadic way of life, give up hunting and instead grow crops? Would they accept being ruled by a king who would collect taxes from them and use these to fund irrigation schemes and an army, as well as a personal lifestyle befitting a king?

If they could have chosen to change their behaviour and organise themselves in this way 50,000 years ago, they would have a good chance of founding an empire that had a lasting impact on human history and evolution. But to do so would mean acting contrary to their most fundamental beliefs about how they should behave as members of their band.

Based on our experience of the few hunter-gatherer tribes that have survived until recently, many of their attitudes, values and moral beliefs would have clashed with the changes needed to progress in evolutionary terms[12]. In hunter-gatherer bands, a man could not be respected if he did not hunt. A male who wanted to plant and tend crops would be despised. The members of other tribes were often seen as sub humans who had to be driven out of the tribe's territory before they stole their game and women. To band together with them would be unthinkable. And anyone who tried to set himself up above the other members of the tribe as a ruler would be seen as a threat to all, and to be stopped at all costs. Anything gathered by an individual was not his or hers, it was the tribes'. Only a deviant would try to accumulate possessions. And deviants were seen as a danger to the band who should be expelled if they did not change their ways.

Such a band would have very little capacity to change its fundamental values and beliefs, and little desire to do so. The members of the band would not have the psychological ability to find motivation and satisfaction in whatever behaviour and life style was needed for future evolutionary success. Merely showing them how things would evolve in the next 50,000 years would not enable them to change their ways. They would continue to find satisfaction in their existing way of life.

But we will see that an organism that develops a fuller understanding of the evolutionary process and of its place in it will be more likely to break free of its biological and social past, and develop the capacity to do whatever is necessary for future success. Such an organism will become aware that its existing beliefs, motivations and values have no special validity. It knows that if its past evolutionary needs were different, its motivations and values would also be different. These predispositions will be seen as the products of shortsighted evolutionary mechanisms that have been incapable of producing the motivations and values needed for future evolutionary success.

The organism will know that all organisms that develop the capacity to mentally model their possible evolutionary futures face a common challenge: to find motivation and satisfaction in whatever actions and behaviours are shown by their models to bring future evolutionary success. The challenge is not only to see what is needed for future evolutionary success, but also to be able to do it. Where necessary, they must cease to serve the beliefs, values and objectives established by their evolutionary past. They must develop the psychological capacity to change their nature. They must be able to change as much as the first cells had to change to produce multicellular organisms. And they must be able to do this not just once, twice, or three times, but whenever necessary.

If they can develop this psychological capacity to adapt their behaviour in whatever way is necessary, they can transcend their biological and social past. They can become self-evolving beings, able to change their behaviour and objectives by conscious choice. They will see themselves as evolutionary work-in-progress, with no fixed characteristics, able to find satisfaction and motivation in doing whatever they choose.

The organism will also know that only organisms that choose to struggle to develop this psychological capacity are likely to make a significant contribution to the future evolution of life in the universe. Those who choose instead to continue to serve obsolete values and motivations will be irrelevant to life, and face eventual extinction. The organism will know that the choice that faces it is, in an evolutionary sense, a choice to be or not to be.

In Chapters 11 and 12 we will look at how organisms such as ourselves are likely to develop this psychological capacity. We will see that an organism can use its modelling capacity not only to model and manage its external environment, but also to model and manage its internal adaptive processes. It can develop mental models of the pre-existing physical, emotional, and mental adaptive processes that determine how it behaves and acts. The models will enable it to understand consciously how its pre-existing adaptive processes operate, what useful effects they have, how they might be modified, and what the consequences of this might be. Through self-knowledge they will develop the capacity to gain control over their internal adaptive processes. Increasingly, this will enable them to manage their physical actions, emotional and motivational states, and their beliefs and other mental processes in whatever ways are necessary to ensure they can do what is required for future evolutionary success. They will develop a capacity for self-management that enables them to revise and modify their previous motivations, beliefs and objectives. These will be revised and managed so that they support the ultimate objective of future evolutionary success[13].

Part Four of the book (Chapters 13 to 19 inclusive) use the ideas about evolutionary progress developed in earlier Chapters to understand the evolution of life on earth. These Chapters identify key evolutionary milestones since life emerged on this planet 3,500 million years ago, and predict important future milestones. A major focus is how human cooperative organisation has evolved, and how it is likely to evolve in the future.

Cooperation amongst hu mans has expanded considerably in scale over the past 100,000 years. Initially cooperation existed only within small family groups. Since then, cooperative organisation has progressively expanded in scale to produce multi-family bands, tribes, agricultural communities, cities, empires, nation states, and now some forms of economic and social organisation that span the globe[14].

We will see that not only has the scale of cooperative organisation expanded rapidly, but so has its evolvability. Human societies have got better at discovering and supporting more effective cooperation, and at adapting it as circumstances change. Modern human societies can adapt and evolve continuously through internal processes during their life. They are not limited to evolving through competition and natural selection between societies.

But the ability of human cooperative organisation to exploit the benefits of cooperation can be greatly improved. Modern human societies are obviously not an endpoint of evolution. The organisms that play a significant role in the evolution of life in the universe will not be those that stop evolving when they reach the position we have. Guided by awareness of evolution's arrow, they will go on to form cooperative organisations of larger and larger scale and of greater and greater evolvability. First they will form a unified planetary organisation that manages the matter, energy and living processes of the planet. Then this organisation will be progressively expanded to form still larger-scale societies of increasing evolvability. Matter, energy and life will be managed on the scale of the organism's solar system and, eventually, its galaxy. The greater the scale of the resources the organism is able to manage, the more likely it will bc ablc to adapt to whatever challenges it faces in its conscious pursuit of future evolutionary success.

We will look at how modern human societies could be changed to improve their ability to organise cooperation to satisfy the needs of their members. Economic markets and governments are the main processes in current societies that support and adapt large-scale cooperation. We will see how these processes could be improved to produce human societies that are more evolvable and better at exploiting the benefits of cooperation. These improvements would establish a highly evolvable and cooperative planetary society. They would produce benefits for all humanity by suppressing conflict and other damaging competition within the society, and by efficiently organising cooperation to serve the needs and objectives of citizens.

But, by themselves, these changes would not establish a society that would consciously pursue future evolutionary success. The society would not achieve the critical evolutionary milestone of using the direction of evolution to guide its future evolution. This is because the society would satisfy the needs and objectives of its citizens, whatever they may be. Until its citizens chose to consciously pursue future evolutionary success, the society would therefore continue to serve only the pre-existing biological and cultural needs of its members. The immense evolutionary potential of a society that could intelligently manage matter, energy and life on the scale of the planet would be used to serve values and objectives established by shortsighted and flawed evolutionary mechanisms. The enormous power of our emerging technologies such as artificial intelligence and genetic engineering would not be harnessed to achieve future evolutionary success. Instead they would be used merely to satisfy obsolete desires and values that conflict with future evolutionary needs.

But this would all change once humans become aware of the direction of evolution and develop the capacity to use it to guide their own future evolution. As humans begin to pursue future evolutionary success consciously and learn how to align their personal values with this objective, they would produce a planetary society that also pursued evolutionary ends. Because the planetary society would manage matter, energy and life to serve the needs and values of its members, it would serve their evolutionary objectives. The society as a whole would develop plans, strategies, projects and goals designed to maximise its contribution to the successful evolution of life in the universe. And it would organise itself to reward and support actions of its members that assisted the society to achieve its goals—just as our bodies reward and support the actions of individual cells that contribute to meeting the body's adaptive objectives.

Finally, we will look at what all this means for each of us as individuals, here and now. We will see that a full understanding of evolution and its direction leaves an individual with very limited choices. It is not open to us to choose to ignore the dictates of evolution. Whether we choose to pursue only the values and motivations established in us by our biological and cultural past, or instead decide consciously to serve the future evolutionary interests of humanity, we will be following evolutionary objectives. The only choice is between serving goals established in us by evolutionary mechanisms that are incompetent, or by mechanisms that are the best available. We can choose to live a life that serves obsolete evolutionary goals established by inferior and shortsighted evolutionary mechanisms. Or we can use awareness of the direction of evolution to guide how we can consciously contribute to the future evolutionary success of humanity.

Once individuals become aware of the direction of evolution, if they decide to continue to serve the dictates of past evolution they are choosing evolutionary failure, in the full knowledge that they are doing so. Individuals that make such a decision are choosing a life that is meaningless, absurd and ridiculous from an evolutionary perspective, and know that they are making such a choice.

Individuals who instead use the direction of evolution to guide their actions obtain a clear answer to one of the most central questions of their existence: "What should I do with my life?" They see that they should do what they can to promote the awareness of the direction of evolution amongst others and to develop in themselves and in others the psychological capacity to do what is necessary for future evolutionary success. They will also want to contribute to the formation of a cooperative and evolvable planetary society that manages the matter, energy and living processes of the planet to form organisations of yet larger scale and of greater evolvability. And they will be aware that their actions are contributing to the next great step in the evolution of life on earth.

One of the most important steps in the evolution of life on any planet is the emergence of organisms who are conscious of the direction of evolution and who use this to guide their own evolution. The actions of individuals who are living now can help ensure that the organism that achieves this milestone on earth will be us.

2

The Causes of Progress

What do I have to do to show that evolution is progressive? If I am to prove that evolution continually improves living processes in some general direction, what must I show about the evolutionary process?

One way to attack these questions is to examine a process that is clearly progressive to discover the type of mechanism that drives progress. Then we can see if the process of evolution also contains such a mechanism. If it does, we may be able to use an understanding of the mechanism to see where it is taking the evolution of life.

One of the areas that shows obvious progressive change is human technology. Technologies that serve key areas of human need such as communication, transport, and lighting have shown long trends of improvement in performance. Completely new and improved technologies have repeatedly emerged in each of these areas. Electric lighting is more efficient and convenient than gas lighting, which in turn is cleaner and more effective that lamps that burn mineral oil or animal fats. Aircraft are more efficient at some transport tasks than railways, which in turn have proved more useful than vehicles hauled by animals. And each century has seen new ways for humans to die in war.

As well as through the discovery of completely new technologies, technology also progresses incrementally. Each successful technology usually shows long sequences of improvement in performance as it is developed. The motorcar of the 1990's is faster, more reliable, and relatively cheaper than cars of the 1920's.

The case that technological change can be progressive is overwhelming[1]. Even strong opponents of the progressive view of evolution generally accept that technological change can be progressive[2]. But what is it that causes progressive technological change? What features of the processes that cause technological change are responsible for progressive improvement? And does the evolutionary process share these features?

Two features seem to be essential if a process is to produce progressive change. First, it must contain a mechanism that searches for improvements and reproduces any that are discovered. In the case of technological change, it is our mental processes that search for improvements. Individuals and groups are continually looking for new possibilities, trying to invent new technologies or to improve existing ones. Whether any particular discovery is reproduced depends on the extent to which it attracts sufficient interest in the economic or political marketplace. Of all the innovations thrown up by inventive minds, the market selects those that will be reproduced. Technology improves by being better able to satisfy the needs of individuals and groups who have economic or other power in these markets.

Second, a process that searches for innovation will produce progressive change only if there is potential for improvement. This potential must be on-going and not be exhausted immediately by the mechanism that searches for innovation. If the potential improvements can be discovered and implemented right away, they will not drive a sequence of progressive change. Obviously, motorcars would not have been improved throughout this century if there had been no potential for on-going improvements, or if all possible advances and supporting technologies had been discovered soon after cars were first developed.

Progress will occur if the potential improvements are discovered in a series of steps, with each step improving on previous steps, but leaving potential for further improvement that will drive further discovery. This will be the case where improvements necessarily build on previous steps, and are able to be made only after the previous steps have been taken. An example is the development of electronic monitoring and regulatory systems in car engines that had to await the emergence of computer technology, that in turn was not possible before improvements in semiconductors.

The pattern of change will also have this structure where better technologies are more complex, or where their development requires comprehensive and detailed research. The simpler technologies will generally be discovered and implemented earlier than more complex ones. For example, the first motorcars and aeroplanes were far less complex than those we see today, and improved versions that are even more complex are likely in the future.

We can therefore expect progressive change wherever it can be shown that there is a potential for improvement that is on-going in this sense. Provided the mechanism that searches for improvements is smart enough to take advantage of the potential for improvement, and not so smart that it discovers them all at once, progressive change will be produced.

With the benefit of hindsight we can see that technology has progressed where there has been potential for on-going improvements, although the potentials were often not obvious beforehand. It is not so easy to show exactly where there will be technological progress in the future. To do so we would need to identify where there are on-going potentials for technological improvement. The difficulties in doing so are notorious. In the middle of the 19th century, engineers could prove beyond doubt that flight by heavier-than-air machines is impossible. Nevertheless, assessments of where there is potential for improvement are made as a matter of course in the planning of research strategies. And where the potential exists, progress can be reasonably predicted.

Other processes will also produce progressive change if they have these two key features—a search mechanism that discovers and reproduces improvements, and potential for on-going improvement. For example, the processes that develop skills in children often have both these features. Children search for new ways of doing things that will improve their skills, often with the help of parents and teachers. Changes are reproduced when they rewarded by success. And it is obvious to all of us who have previously mastered new skills that there is often an ascending scale of potential improvements that will drive progressive change. As a result, children progressively improve their skills as they grow and learn.

This is particularly clear in learning a musical skill such as piano playing, where each improvement necessarily builds on previous ones. There is a sequence of potential improvements in playing ability, with each improvement building on the skills gained earlier. We therefore can predict confidently that if a child has the will and ability to learn (i.e. an adequate search mechanism) progressive improvement will follow.

Unless a process contains both of these key features in full, change will not be progressive. For example, there is obvious potential for on-going improvement in the ability of human society to satisfy the needs and to develop the potential of its members. Technological progress has opened up the potential to improve the living standards of all people on this planet. But this potential for social improvement has not been fully exploited this century. It has not driven on-going social progress. Instead, at the end of the 20th century over 800 million people are chronically undernourished, and each year nearly 13 million under fives die as a direct or indirect result of hunger and malnutrition[3]. In industrialised countries, many people spend much of their lives in unsatisfying, boring and meaningless work, while large numbers choose drugs over reality.

Human society has failed to produce on-going social progress in these areas because it does not include a mechanism that selects and reproduces social improvements when they are discovered or proposed. If society stumbles on improvements either by accident or by conscious effort, there is no process that locks in the improvements and perpetuates them. The economic and political markets that select and reproduce technological advances do not do the same for social improvements.

As we shall consider in detail later in this book, it is possible to organise human society so that social improvements are selected and reproduced, and so that social progress is as inevitable, unstoppable and natural as technological progress. And we will see that evolutionary success for humanity will ultimately depend on organising ourselves in this way.

This discussion has now got us to the point where we know what must be done to show that evolution is progressive. First we must show that the evolutionary process contains mechanisms with the ability to search for improvements in living processes and to reproduce any that are discovered. Second we must show that there are potentials for improvement in living processes that are on-going.

It seems certain that any process that successfully evolves living processes must meet the first condition. The central feature of an evolutionary process is that it searches for useful changes in living things, and reproduces improvements from generation to generation. If a process is capable of producing evolutionary change in living processes, it must contain a mechanism that searches for improvements and reproduces any that are discovered.

The most familiar evolutionary mechanism, natural selection acting on genetic variation, clearly meets this condition. Animals produce young that may vary genetically from their parents because of mutations or different combinations of genes. If the variants prove to be better adapted because of their different genes, they are likely to have relatively more offspring, with the new genes eventually taking over the population. In short, this process searches for improvements by throwing up genetic variants, and then perpetuates any improvements that are discovered.

The fact that the genetic mechanism searches for improvements largely by blind trial-and-error does not prevent it from producing progressive evolution. If there are potentials for on-going improvement and if it can stumble on them, it will produce progressive change. It does not matter that the genetic mechanism has no foresight or intention to make progressive discoveries.

Although nearly all living organisms known to man evolve through the genetic mechanism, it is not the only mechanism that produces evolution. There are other evolutionary mechanisms that search for better adaptations in living processes, and reproduce the improvements from generation to generation. A number of the processes which adapt humans as individuals provide obvious examples. As individuals, we continually use our mental processes to search for better ways of adapting to our circumstances, better ways to stay healthy, to prolong our lives, and to satisfy our needs. Often assisted by others, we develop and test possible adaptations mentally, select those that we think will work, try them out in practice, and adopt those that are best for us. When a successful improvement is discovered, it is likely to be adopted by others. If so, it will be reproduced throughout the population. It will be preserved across the generations until it to is replaced by something better. The result is an evolving culture of adaptive knowledge.

Change produced in this way is as much evolution as is genetic change produced by natural selection[4]. When humans discovered how to build aircraft, and passed the ability from generation to generation, the result was evolutionary change. It was as much evolutionary change as the discovery by natural selection of wings in the first birds, and the passing of the relevant genes from generation to generation. The mechanisms that discovered and reproduced each of these types of change were different, but the end result was the same: evolutionary adaptation.

In humans, the processes that adapt individuals are not the only evolutionary mechanisms. Other mechanisms adapt human societies and organisations. The processes that adapt the cells within our bodies also fall into two classes. One class adapts individual cells, and the other adapts the organisations of cells that form tissues, organs, and the body as a whole. And in ant colonies, some processes adapt individual ants, and supra-individual processes adapt groups of ants and the colony as a whole.

In the case of the mechanisms that adapt individuals, possible improvements are developed, tried out, and selected at the level of the individual human, cell or ant. But in the case of organisational adaptation, it is the collective activity of many individual humans, cells or ants that search for and reproduce adaptations. Through these processes the organisation can solve adaptive problems collectively that individuals cannot. Just as our brain solves problems that are not understood by any cell in the brain, human markets and governments can solve adaptive problems that are not fully understood by any individual.

We will deal in detail later with the evolution of these various evolutionary mechanisms. We will look at what has driven the evolution of new mechanisms and how each mechanism has evolved and got smarter at evolving. Then we will consider how the mechanisms that currently evolve human individuals and human societies will evolve and improve in the future.

However, what is clear already is that the processes that produce evolutionary change each include a search mechanism. But this by itself is not enough to prove that evolution is progressive. We must also show that there are potentials for improvement in living things that are on-going and that can be exploited by the search mechanisms.

Are there on-going potentials for improvement in living things? Will an evolutionary search mechanism that is smart enough discover a sequence of changes, each better than the one before, and each an improvement in purely evolutionary terms?

Once it is accepted that evolution includes adaptive change in human culture and society, we can immediately point to an area where potentials for on-going improvement exist and have driven progressive evolutionary change: human technology. As we have seen, technological change is often progressive, producing sequences of improvements, each step better at satisfying human needs than those before. And it can be shown that there is further on-going potential for improvement in many areas of technology that will drive further progressive change.

We can go a step further by noting the general similarity between human technology and the bodily features of organisms that adapt them to their external environment. We often use our technology to adapt to our external environment. But most organisms achieve this through specialised features of their bodies such as legs, wings, fins, eyes, gills and lungs. The technology that enables them to move, see and communicate is part of their bodies.

Because of this general similarity, we can expect that the potentials for on-going improvement that have driven technological progress will also exist for the bodily technology that adapts organisms to their environment. And our experience with technological progress points to where these on-going potentials will be most pronounced. They will be most obvious in the evolution of complex adaptations such as the eye where a number of components need to cooperate together to make the adaptation effective. Evolution is unlikely to be able to perfect these complex adaptations in one step. Instead we would expect that the adaptation would be improved in a series of steps as new components are added and developed, and as components are adjusted to changes in other components[5].

These sequences can be expected to be most noticeable where the improvements are chasing better adaptation to environmental conditions (living or non-living) that are themselves changing. The most obvious examples are 'arms races' between predators and their prey: the predator evolves greater speed, the prey counters by increasing its evasive ability, the predator responds with improvements in its ability to anticipate the movements of the prey, and so on[6].

The fossil record, where it is good enough, confirms the expected progressive evolution of complex adaptations. A particularly clear case is the evolution of warm-bloodedness in the ancestors of mammals. An ability to continually maintain the warm body temperatures that are best for movement and other activity has significant advantages. Cold-blooded reptiles are unable to do this, so they need to wait for warm air temperatures and the sun if they are to get to a temperature that is best for active movement. Early warm-blooded mammals could be out feeding all night while cold-blooded reptiles waited for the sun.

Complex changes to the blood system, physiology, and metabolism were necessary for efficient warm-bloodedness. Amongst the key changes needed were the evolution of external hair to retain heat, and the development of a four-chambered heart to improve blood circulation. The history of the evolution of the early mammals reconstructed from the fossil record shows how these various components were improved individually and as a combination over a long period[7]. This progressively enhanced the ability of mammals to maintain a warm body temperature no matter what the external temperature.

So we have good reason to believe that much of the technology embodied in the adaptations of organisms will show the same potential for on-going improvement as many areas of our external technology. And the search mechanism of natural selection acting on genetic variation will progressively discover these potentials, producing progressive evolution in particular adaptations until the potential for improvement is finally exhausted.

Although the potential for progress in external technology and in the complex adaptations of organisms is similar, the actual rate of progress and its directness is very different. And these differences go a long way toward explaining why progress in genetic evolution is not as widely recognised as technological progress. Although both processes are fundamentally progressive, technological progress is a lot easier for us to recognise because it is more direct and it is faster, occurring during our lives.

The reason for these differences is that the genetic search mechanism is far inferior to the mechanisms that develop our technology. Genetic evolution is not very smart at exploiting the potentials for on-going improvement. Unlike us when we try to improve technology, it has no capacity to plan ahead, to use foresight, to visualise possible improvements, or imagine alternatives and try them out in its head before making them. Instead, without any idea of what might work and where it might lead, the genetic search mechanism blindly changes adaptations when new organisms are produced, and sees how the changes work in practice. There is nothing to stop the genetic search moving in the opposite direction to where the greatest improvements lie. A genetic change will be favoured by natural selection if it is better than what has come before, even though it might move away from greater potential improvement. And a change that is a step towards greater potential improvements will be reproduced only if it is itself an improvement.

Genetic evolution operates a bit like trying to improve a 1930's motorcar by making random changes to it in the dark, without any knowledge of how the car works or how it might be improved, and using parts you don't understand. The only feedback you would get about how you are going is to be told when a particular change is an improvement. We know from the history of cars since the 1930's that there are many on-going sequences of potential improvements that you could discover. However, if you set out to discover improvements using only blind trial-and-error, it would take you a very long time to make much progress and when you did, it would be very indirect.

Even where your search mechanism eventually exploited some of the potentials for improvement, the history of the changes that got you there would not look very progressive. There would be periods in which the search got stuck and no improvements were made, periods in which only minor enhancements were discovered and much better improvements were missed, and periods in which the changes that were made moved in the opposite direction to the potential for greatest progress. If you tried to decide whether such a process was fundamentally progressive by looking only at the history of changes, it would be very easy to get it wrong. It would be easy to get lost in the trees, and never see the wood.

This is the way that genetic evolution proceeds, even where the existence of potentials for on-going improvement makes it fundamentally progressive. The evolution of birds provides an example. It is evident that with the emergence and rise of land insects, large flying organisms that fed on the insects could be successful. Eventually this potential role was filled by birds. Despite the potential for success, the genetic mechanism was unable to evolve a group of insects to take up this role, probably because the insects were locked into having a hard external skeleton that is not as efficient for larger organisms as an internal skeleton. Although the insects were already there on land, it was not until many millions of years later that evolution eventually filled the role, and then only by an extraordinarily circuitous path. The ancestors of the birds came not from the land but from the water as amphibians.

The genetic evolutionary mechanism was completely unable to get itself unstuck by developing an insect with an internal skeleton. If it had done so, the role could have been filled quickly and directly by large flying insects. But in the absence of an asteroid or other catastrophe to undo enough of the evolution that had got it stuck, the genetic mechanism could not directly and immediately exploit the potential role.

So the opponents of a progressive view of evolution have found it very easy to go to the history of life on earth and point to absences of progress over long periods[8]. But they have failed to see that extensive periods without progress are an inevitable consequence of a blind trial-

and-error genetic search mechanism. The survival of some groups of animals without significant improvement for millions or even billions of years is not conclusive evidence against progressive evolution. There may be potential improvements in those groups that genetic evolution has not yet been able to discover. Life may have to await the evolution of smarter evolutionary search mechanisms before evolution can explore some of these potentials.

Whether or not evolution is progressive, there will be species that do not progress for very long periods of time. To prove the case against evolutionary progress, it is not enough to point to organisms like bacteria that have flourished unchanged on this planet for more than 3,500 million years. Anti-progressionists must also show that there are no potentials for on-going improvement that the evolutionary search mechanisms have failed to exploit. They must show that there can be no alternatives to current bacteria which, if eventually produced by evolutionary search mechanisms, would prove far more successful than current bacteria in strictly evolutionary terms. They must show that, for what they do, bacteria are perfect, and there can be nothing better. And they must show that this will remain true whatever happens in the future evolution of life in the universe.

Gould and other anti-progressionists have not even attempted to meet this critical challenge. They have not explored the possibility that there are potentials for on-going improvement that evolutionary search mechanisms have failed to discover. They have not considered whether future evolutionary developments are likely to overtake bacteria in their current form.

A major task of this book is to take up this challenge. I will show that there are general potentials for on-going improvement that will drive future progressive evolution. And I will be showing that bacteria (and humans) that don't explore these potentials for progressive improvement will not participate successfully in the future evolution of life in the universe. They will progress or perish. We will progress or perish.

So far in this Chapter we have seen that there are likely to be potentials for on-going improvement in the technology embodied in the complex adaptations of organisms. Wherever the use of genetic trial-and-error is able to exploit these potentials, progressive evolution of the adaptations can be expected. But will this evolution produce a general advance in living processes? Will life as a whole progress in evolutionary terms? Or, as argued by Gould and his supporters, will this evolution produce only better adaptation to local conditions? Will progressive change occur only while there is room for improvement in adaptation to the specific and localised environmental problems met by each population of organisms?

We would expect to find general progress if all organisms had similar potentials for on-going improvement, no matter what the nature of their local environmental conditions. The general potentials would drive general advance. In what circumstances might we expect to find general potentials of this kind? Potentials would be general if a particular adaptive problem is encountered by all organisms. We would expect general progress in the development of the adaptive technology to deal with the problem. General potentials for improvement would also exist if there were complex adaptations that would improve all organisms, no matter what their local environment.

Where these general potentials exist, problems and opportunities faced locally by populations would also be problems and opportunities faced generally by all other populations of organisms. So if a population were to exploit some of these general potentials for improvement, it would not only improve its adaptation to its local environment, but also would improve in a general sense. It would improve in all environments. By achieving better adaptation to its local conditions, the population would participate in universal advance or progress.

But do these general opportunities and problems exist? It is certainly the case that some adaptive challenges and opportunities are very general, affecting many species. The need for some form of vision is widespread, as are the benefits of warm-bloodedness. And when we examine the history of life on earth, we find general advances in particular adaptations across many species where common adaptive problems or opportunities are found.

For example, eyes have evolved in many different organisms, as have the various features that were needed for warm-bloodedness in mammals. There are also many examples of general improvements that evolved in one or a small number of closely related species and then spread widely because they brought the species general evolutionary success. The general nature of these improvements enabled the original species to adaptively radiate into many other environments, often ousting species that had not developed the general improvements. The fossil record shows many spectacular adaptive radiations triggered by the discovery of complex general improvements. For example, major radiations followed the emergence of the first primitive fishes in the Devonian, the evolution of the first primitive reptiles in the Permian, and the rise of the first dinosaur-like reptiles in the Jurassic[9].

Of course, even where improvements provided general advantages in this way, many species were not replaced immediately by those with the improvements. In the case of some 'living fossils' they have still not yet been replaced. A species would not be ousted where it had valuable and specialised adaptations to local conditions that were not outweighed by the general improvements of the more progressive invaders.

This is another example of where the limitations of the genetic evolutionary mechanism mean that genetic evolution does not progress as quickly or as directly as technological evolution. When a general technological improvement is discovered, such as a lighter but stronger alloy, it can be readily combined with any other existing technology to produce improved solutions to particular problems. Technological evolution can easily combine the best with the best.

The genetic mechanism is unable to do this. It cannot directly combine adaptations discovered in different species. When a relatively general advance such as a better eye is discovered, there is no process that enables it to be immediately adopted by all those species that have inferior eyes. Again we find that although genetic evolution is as fundamentally progressive as technological evolution due to the existence of similar potentials for on-going improvement, it is far inferior at exploiting these potentials.

It is clear that there are adaptive challenges and opportunities that are not purely local. Some are general in the sense that they affect many species. But are there potentials for on-going improvement that affect all living processes, that are universal, and that will therefore drive the progressive evolution of life as a whole?

As I indicated in the first Chapter, I will be arguing that evolution progresses towards increasing cooperation amongst living things. To succeed with this argument I have to demonstrate that increased cooperation provides a general potential for on-going evolutionary improvement. To demonstrate this, I will have to establish a number of things. First, that increased cooperation amongst living processes has the potential to provide greater evolutionary success. Second, that this principle is general and the potential for greater evolutionary success applies to all living processes. Third, that the potential for improved evolutionary success is on-going, because evolution is unable to immediately exhaust the potential benefits of increased cooperation.

I will show that evolution exploits the benefits of cooperation amongst living entities through the formation of complex organisations of those entities. The organisations are structured so that cooperation is supported within the organisation. On this planet, evolution has produced cooperative organisations of molecular processes to form cells, cooperative organisations of cells to form multicellular organisms such as insects, frogs and ourselves, and cooperative organisations of humans to form human societies.

However, the formation of these structured organisations enables cooperation to be organised only between the living entities *within* the organisations. Evolution will not be able to exploit immediately the benefits of cooperation *between* the organisations. These benefits will not be exploited until the organisations themselves are organised into organisations, producing cooperative organisations of organisations. For example, the organisation of molecular processes into cells produced cooperation between molecular processes. But the benefits of cooperation between cells were not exploited until cells were organised into multicellular organisms. And cooperation between multicellular organisms was not exploited until the evolution of societies of organisms.

Continued repetitions of this process forms cooperative organisations of larger and larger scale, each containing the smaller-scale organisations that have evolved previously. As a result, human social systems include humans which include cells which include molecular processes. This evolution of organisations of larger and larger scale extends the scale over which living processes are organised cooperatively, but leaves unexhausted the potential for cooperation between organisations of the largest scale. The potential for further beneficial cooperation will not be finally exhausted until all living processes are permanently organised into a single entity that is of the largest possible scale. The potential for increases in the scale of cooperation in this universe will end only when the entire universe is subsumed in a single, unified cooperative organisation of living processes. It will end only when the matter, energy and living processes of the universe are managed into a super organism on the scale of the universe.

This evolutionary sequence has all the features of a process that is fundamentally progressive. The potential for almost indefinite expansion in the scale of cooperative organisation and the advantages this will bring provides a potential for evolutionary improvement that is on-going. And each step in the sequence of improvements builds on and goes beyond the improvements made in the previous steps.

The next five chapters are devoted to the development of these arguments in detail. We begin in Chapter 3 by exploring how cooperation can benefit living processes in evolutionary terms. We will see that by cooperating together, living processes improve their ability to meet whatever evolutionary challenges they face. Increased cooperation has a general potential to provide greater evolutionary success. And the wider the scale over which the cooperation is organised, the more successful the cooperators can be.

PART 2

The Evolution of Cooperation

3

Why Cooperate?

Without cooperation, you would not exist.

Almost everything we use and depend on in our everyday life is produced and brought to us by the coordinated actions of many other people. Almost everything made by humans is produced cooperatively.

Consider a very simple manufactured device such as a ballpoint pen. It is the product of the coordinated work of thousands of people across the planet: manufacturing and moulding the plastic components, collecting and processing the petrochemicals that are used to make the plastics, manufacturing the metal components, producing the alloys from which these are made, mining and processing the various ores used in making the alloys, transporting all these raw materials and components around, inventing, designing and refining the design of the pen, and so on. None of these tasks are carried out by individuals acting alone. They are done by individuals who are members of cooperative organisations such as multi-national companies. These organisations in turn use materials and services produced by other cooperative organisations and so on. Everything is the product of an immense network of cooperation.

Even unprocessed food such as vegetables are produced through the cooperative activities of many people: fertilizers, insecticides, farm equipment, irrigation systems, transport, and so on, are all produced and organised cooperatively. And each individual involved in this production relies on many other goods and services that are also generated cooperatively.

Not only is our physical life completely dependent on cooperation, so too is our mental life. Many people have contributed to the development of the ideas in our culture about how the world works, the place of humanity in it, what is right and wrong, how we should live our lives, and how we should relate to others. Without exposure to these ideas we could not think as we do and we would not have our current worldviews. What we think and the contents of our minds is as much the product of social cooperation as our technology and our food.

Viewed from afar, human society is a dynamic network of cooperative activity that inseparably interlinks our lives and our actions. But it is not just that we are all totally dependent on the cooperation of those around us. We are also totally dependent on cooperation within us. We are composed of cooperative living processes. If the living processes that make up our bodies did not cooperate, we would not exist. Their cooperation is us.

Each of us is an organisation of about a million billion cells. These cells are specialised into many different types that team up to form organs such as the heart, stomach, and bones. The digestion of food, the transport of food to cells, the fight against invading cells that cause disease, the transport of oxygen to cells and the use of thought to solve problems are all the product of extensive cooperative amongst highly specialised and differentiated organs and cells[1]. Like the goods and services produced in human society, the key functions of the human body are all the product of the coordinated activities of thousand and thousands of differently-specialised and interdependent cells. If this cooperation breaks down, so to do our bodies. Cancer is one example of what happens when cells go their own way at the expense of the body.

And the cells themselves are cooperative organisations. Without extensive cooperation between the molecular processes and organelles that make up cells, we would not exist. Each of our million billion cells is made up of thousands of incredibly small and intricate parts that cooperate together to produce the functions of the cell. At the molecular level, a human cell may contain up to 100,000 different proteins. Proteins are differentiated and specialised in many ways: some provide support for the components of the cell, some form part of the cell membrane, some have the ability to contract, moving the cell or parts within it, and many are specific enzymes that regulate the essential chemical reactions that enable the cell to function. Often these enzymes cooperate to form teams in which each member regulates a particular step in a sequence of reactions.

Just as humans team up to form corporations, and cells are organised into organs, groups of molecules may also form larger structures called organelles. These too are differentiated and specialised, with mitochondria, the so-called power houses of the cell, releasing and converting energy for use in the cell, ribosomes providing sites for putting together proteins, the nucleus housing most of the genetic material, and lysosomes providing places in which food can be digested without dissolving other parts of the cell[2].

As with our bodies and our social systems, the functions of our cells are produced by the cooperative activities of many specialised and differentiated components.

In summary, we are cooperators that are made of cooperators that are made of cooperators and so on. It is cooperation all the way down.

But why all this cooperation? What are its advantages? Are they the sort of advantages that can drive evolutionary progress? In Chapter 2 we saw that evolution would be progressive if there were general potentials for improvement in living processes that were on-going. Are the potential benefits of cooperation on-going in this sense? As evolution proceeds to exploit the advantages of cooperation, will there always be potentials for further improvement that will draw evolution forever onwards?

We know that much of the cooperation we see around us must be advantageous because it has succeeded in competition with alternatives. Cooperation is widespread only because it has been able to defeat non-cooperation in evolutionary struggles. The teams of cells that have formed multicellular organisms such as birds, insects and mammals have proven competitively superior to individual cells in many environments. And the teams of molecular processes that have formed cells have almost completely replaced the less cooperative molecular processes that preceded them.

But why are these cooperative organisations competitively superior? Why can a group of living entities that team up do better at evolution than individuals who do not?

The key to the success of cooperation is that combinations of individuals whose activities are coordinated can do things better than individuals, and can do things that individuals cannot.

One of the main ways in which cooperation enables things to be done better is through specialisation and division of labour. In a cooperative organisation, every individual does not have to do everything needed for survival. Instead, an individual can specialise in a particular task, performing the task not only for itself but also for others in the group. This lets the others specialise in other tasks for the group. The result is a highly interdependent organisation in which key tasks are performed by individuals who are specially adapted and equipped to do them[3].

So we do not all have to do the work of a plumber, a carpenter, and a medical doctor. We do not have to spend time learning all these skills, we can specialise in one job and do it far better than any individual who had to learn them all. And because we do the one job full time, it pays to develop special tools and equipment to enable us to do the job even more efficiently.

Similarly, a specialised nerve cell is free to develop the structure and processes needed for it to transmit electrical impulses effectively. It does not have to retain a structure that enables it to perform all the other tasks of cells in our bodies. And the cooperative division of labour within cells enables protein enzymes to specialise in particular tasks. So an enzyme can develop the structure needed to regulate one specific step in a chain of reactions, rather that having to have a generalised structure that can control many steps.

Sometimes individuals also have natural advantages that make it more efficient for them to specialise in particular tasks. For example, cells at the front of animals are best placed to sense the environment because they meet new conditions first. And many Portuguese farmers specialise in wine production because of the suitability of Portugal's climate.

In all these ways, a cooperative division of labour can enable living processes to function more efficiently and effectively. It can improve their ability to get food, to move, to fight enemies, to solve problems, to understand how the world works, to use this knowledge for better adaptation, and to evolve. Whatever evolutionary challenges face a group of individuals, they can deal with them better using a cooperative division of labour.

So wherever evolution has been able to fully exploit the benefits of cooperation, we always find the extraordinary level of specialisation and interdependency that results from a high degree of division of labour. We find it within cells, within our bodies, within our social systems, and between nations[4]. And there is every reason to believe it will also be a feature of organisations that are capable of future evolutionary success on even larger scales.

But it is not only through a division of labour that cooperation provides advantages. Cooperation is also able to exploit the fact that combinations often have new features that their components do not[5]. Combinations can do things that individuals cannot. We are most familiar with this in non-living things: nickel and chromium combined with steel produce stainless steel. Unlike steel alone, it is rust and tarnish resistant. And bricks combined together in various ways can make a house or a bridge or some other useful structure. Individual bricks cannot.

Similarly, combinations of living things can also do things that individuals cannot: amino acids, the building blocks of proteins, are unable to regulate chemical reactions in the cell by themselves. But when combined to form protein enzymes, they can control and manipulate other atoms and molecules, determining how they react and what larger molecules are constructed in the cell. A flock of birds or a troop of baboons is able to detect stalking predators that individuals are unable to. And combinations of cells in our bodies form tubes to carry blood, teeth that enable us to chew food, and many other useful structures that individual cells cannot.

Cooperative combinations can also have significant evolutionary advantages because of their larger scale. For example, larger-scale organisms can have adaptations that are more complex. A bacterium could not evolve a brain as complex as ours. This level of complexity was possible only after billions of cells teamed up to form multicellular organisms. Larger scale can also provide power over smaller groups, over individuals of the same species and over individuals of other species that are used for food. Larger-scale human groups are generally better at defending and taking territory. Many animals combine in groups to chase off predators that would easily overpower a single individual. Predators such as lions and dogs that combine to hunt as groups are able to round up and kill larger and faster prey than they could as individuals. And males in a number of species including dolphins and baboons often team up to successfully fight for sole access to females[6].

But more importantly, the ability to form larger-scale cooperative groups enables organisms to manage and manipulate their environment over larger scales. Large human organisations can operate mines, build dams and establish communications networks of a scale unimaginable to small bands of earlier humans. A human social system organised on the scale of the planet is likely to be able to develop the capacity to prevent large asteroids from colliding with the planet. Dinosaurs and bacteria have been unable to adapt on a sufficient scale to do this. In general, multicellular organisms can cope successfully with larger-scale environmental threats than can single celled organisms. And single cells can adapt to larger-scale threats than could the first living molecular processes.

A further very general advantage of cooperation is that it can prevent the harmful effects of destructive competition. Non-cooperating individuals pursue their own interests even where this damages the interests of others[7]. A population of such individuals will damage each other's interests, and all will loose. Obvious examples are fighting between animals over food and territories and the chemical warfare waged between plants over places to live. Competition between rabbits on an island can result in the destruction of all food sources, resulting in the death of the rabbit population. Pollution is another example. A company that seeks only to maximise its profits will engage in activities that degrade the environment if this produces the highest profit.

The advantages of cooperation mean that a whole world of new adaptive opportunities is opened whenever living processes team up to form a cooperative organisation. This has clearly been the case when molecular processes teamed up to form cells, when cells teamed up to form multicellular organisms, and when humans teamed up to form social systems. Each of these levels of cooperative organisation has been able to conquer new environments and develop adaptations unknown to the levels that preceded them. We can be sure that new evolutionary opportunities will also be opened up as human organisation expands to the scale of the planet and beyond. Exactly what these new opportunities will be, we cannot know with certainty. The adaptive opportunities that were opened up by the move from single cells to multicellular organisms are unimaginable from the point of view of a single cell and its life experiences. What could the trunk of an elephant, the wings of a bird, the heart of a deer, or a jet aircraft possibly mean to a cell? But what we can know with certainty is that by expanding the scale of human cooperative organisation we will open up greater adaptive capabilities. And as we shall see in later chapters, it will increase the chances that humanity can play a significant role in the future evolution of life in the universe.

The evolutionary advantages of cooperation are significant. Wherever evolution is able to exploit these advantages by organising cooperation, it will do so. There are two key reasons why the evolution driven by these advantages is likely to be progressive: first, the advantages of cooperation are general. They apply to cooperation between any living processes. They do not depend on the existence of any special local circumstances or conditions. Any organisms, whether of the same species or not, can benefit from the evolution of suitable cooperative relationships between them. Whatever the evolutionary challenges faced by organisms, they can meet them more effectively through cooperation.

This brings us to the second reason why the advantages of cooperation can drive progressive evolution. They can do so because the advantages continue to apply no matter how large cooperative organisation becomes. The advantages do not cease once cooperative organisation reaches a particular scale. Further increases in cooperation will deliver further evolutionary advantages. Increases in the scale of cooperative organisation did not stop providing advantages once cooperation reached the scale of a single cell, or the scale of multicellular organisms, or of human villages. In all these cases, the potential benefits of cooperation between organisations of the largest scale continued to drive the progressive evolution of cooperation.

Currently, the potential benefits of increased cooperation are expanding human organisation to the scale of the planet and beyond. There is every reason to believe this expansion will continue. No matter what the scale of cooperative organisation, greater benefits will be achieved by further increases in scale, whether by the expansion of existing organisations, or by the evolution of cooperation between organisations of the largest scale. Living processes that cooperate over the scale of a solar system will have much greater adaptive capabilities and opportunities than us. But their abilities will be far inferior to organisms that cooperate to manage the matter, energy and living resources of a galaxy. The potential benefits of cooperation can be expected to continue to drive increases in the scale of cooperative organisation at least until the universe is organised into a single cooperative organisation of living processes.

However, the potential benefits of cooperation can drive progressive evolution only to the extent that evolution is capable of exploiting the benefits. If evolution is unable to find ways to make cooperation work, progressive evolution will stall. This is a serious difficulty if the views of many mainstream biologists are correct. Most evolutionary theorists believe that evolution is very poor at organising cooperation, and that self-interested organisms will usually win evolutionary struggles. Consistent with this, there are many instances in the world around us where evolution seems unable to organise cooperation. In the next Chapter we will look at barriers to the evolution of cooperation that have been identified by evolutionary theory, and see what implications the barriers have for progressive evolution.

4

Barriers to Cooperation

If cooperation is so good, why isn't it universal? If cooperative organisations are better in purely evolutionary terms, if they are able to out-compete non-cooperators in evolutionary struggles, why are we surrounded by so many examples of animals that aren't cooperating? Why haven't the potential benefits of cooperation that we discussed in Chapter 3 already driven the evolution of cooperation amongst all living things?

We humans are obviously only partial cooperators. It is true that human economies can be spectacularly cooperative with an extraordinary specialisation and division of labour. It is also clear that internally, within our own bodies, we are cooperative all the way down. But anyone who tries to convince us that humans are always cooperators has a hard job in front of them.

The 20th century has seen millions killed in the largest wars in human history. Hundreds of millions of people in the world are chronically underfed, even though there are sufficient resources to feed us all. Millions die each year as a result[1]. Thousands of millions will live and die without having any chance to obtain the knowledge and skills needed to participate effectively in a modern economy. They will not share in the understanding of the world and of ourselves that has been made possible by the growth of human knowledge. They will live and die without fulfilling the potential for personal development and self-awareness that is possible at this time in the evolution of humanity.

And individuals, corporations and countries continue to pollute and degrade our environment even though they know they are damaging the lives of others, now and into the future.

On a smaller scale, most of us do not feel that we are cooperators. There are strict limits to what we are prepared to give up for the common good. We know that tens of thousands die each year from starvation. But most of us will not sacrifice much to save any of them. We do not do for others what we would have them do for us.

We are very choosy about whom we will cooperate with: we are more comfortable cooperating with people who are likely to return any favours and to share the benefits of cooperation with us. So we cooperate more with family, friends and people we know we can trust. We are far more wary about cooperating with strangers. Before we do so, we want to be sure the strangers are trustworthy, or are put in a position where they will not take more than their fair share of the benefits of cooperation. For example, where the cooperation is a business venture we might insist on an enforceable contract, or that the investments and profits are not solely under the control of the stranger.

The other animals about us cooperate even less than we do. Like us, each of them is a cooperative organisation of smaller-scale living processes. But very few animals cooperate with each other. There is little cooperative organisation amongst the members of most species of earthworms, snails, crabs, prawns, spiders, insects, fish, frogs, snakes, lizards, birds or mammals. There is some cooperation between parents and offspring, but little beyond this. And in the few cases where cooperation is more extensive, it is nowhere as complex as the cooperative division of labour found within the bodies of individuals and within human society. Even ant societies have differentiated into at most four different types of ants.

If cooperation is so good, why is cooperation between animals so scarce and underdeveloped? There is no controversy amongst evolutionists about the answer. Cooperation does not evolve easily. There is a fundamental barrier to the evolution of cooperation. Why is this so?

In most circumstances, the only features that natural selection will produce in animals are those that benefit the individual. No matter how much a feature benefits the group or the species, it will not evolve unless it also benefits the individual. So if a feature causes an animal to help others without any benefit to itself, the feature will die out. And it will die out even if the benefits provided to others far outweigh the costs to the cooperator[2].

The reason for this is simple. The only way the gene that causes an individual to be cooperative can survive in a population is by reproducing successfully. For this to happen, the gene must cause individuals who carry it to have greater numbers of successful offspring than individuals that do not carry it. If the cooperator gene does this, the numbers of individuals in the population that carry the gene will increase, until eventually all will be cooperators. If it does not do this, non-cooperator genes will do better, and cooperator genes will die out.

So what will happen to a cooperator gene that causes its carriers to use resources to help others without benefit to themselves? How will it do in competition with non-cooperators who use the same resources to help themselves rather than to help others? There is no doubt that the cooperator's ability to survive and reproduce will be inferior to non-cooperators, and the cooperator gene will die out.

The more the cooperator sacrifices its own interests to help others, the worse it will do and the quicker it will die out. And cooperator genes will die out no matter how superior the advantages of cooperation. In fact, the more efficient and effective cooperation is at providing extra benefits to others, the worse the cooperator will do. This is because some of the others it helps are likely to be non-cooperators. And the more the cooperator helps its competitors, the worse it will do in comparison.

Evolutionary struggles tend to be won by genes that cause individuals to put their own interests ahead of the interests of others. The useful effects that an individual has on others will not improve its own competitive ability. Only the effects it has on itself will do this. This essentially is the 'selfish gene' perspective that has been argued so persuasively by writers such as Richard Dawkins[3] and George Williams[4].

They have shown that there is a fundamental barrier to the evolution of cooperation amongst animals that can be overcome only in limited circumstances. Even though cooperation can provide significant advantages, the genetic evolutionary mechanism is not very effective at exploiting them. The 'selfish gene' perspective can successfully explain why we see only limited cooperation amongst animals.

The conclusions of the 'selfish gene' perspective are an important part of my demonstration that the potential benefits of cooperation can drive progressive evolution indefinitely. If evolution could rapidly exploit all the benefits of cooperation amongst living processes, there would still have been a period of progress towards increasing cooperation, but it would have been short-lived. The potential benefits of cooperation would soon have been exhausted. The potential would not have been on-going in the way needed to drive indefinite progress. Paradoxically, the reason why cooperation continues to be so important in the evolution of life is that it does not evolve easily.

But as we discussed briefly in Chapter 3, the barrier also presents a direct threat to progressive evolution. Evolution will not progress if the barrier to the evolution of cooperation is insurmountable. This threat is even greater once we recognise that the barrier to the evolution of cooperation does not apply only to the animals we see around us. It is not just genetic evolution that is restricted by the barrier. The 'selfish gene' perspective can easily be extended to show that a similar barrier applies to the evolution of cooperation amongst all other living processes. It applies to cooperation amongst human organisations, individual humans, single cells, and molecular processes[5].

As is the case for the animals around us, evolutionary struggles amongst other living processes will be won by individuals that put their own interests first and foremost. Those that sacrifice their own interests for the interests of others will die out. The advantages of cooperation can be exploited only where evolution finds a way around this barrier. As we shall see in detail later, cooperation emerges only where evolution discovers how to build cooperative organisations out of self-interested components.

A number of examples will illustrate how the barrier to the evolution of cooperation applies to these other living processes.

Consider human corporations that sell their products in a highly competitive market. The corporations that are less efficient will tend to be out-competed, and will go out of business or be taken over. What will happen if the most efficient way corporations can make their products also pollutes the environment? Any corporation that cooperates with the community by reducing pollution will be less efficient and will go out of business. The corporations that pursue only their own interests will be more competitive, and will pollute. No matter how good the intentions of a corporation, it cannot stop polluting if it is to survive. And it does not matter whether it is run by responsible people who genuinely care about the environment. All they will achieve if they steer the corporation towards environmental responsibility is to send it broke.

The so-called "free rider problem" undermines cooperation in many human activities[6]. For example, it prevents businesses in an industry from cooperating together to train sufficient employees for the industry. If an industry is to be successful and to expand, enough workers must be trained in the general skills needed in that industry. However, businesses that make the investment needed to train employees can have their trained employees poached by other companies. This will often happen before the businesses have got a good return on their investment in training.

Free rider companies will rely on other businesses to train the skilled employees they need. Free riders will end up in front because they can get trained employees without paying the high costs of training. In contrast, companies that train can end up having paid for training without being able to hold onto trained employees. To remain competitive, more and more companies have to reduce their general training as much as they can, and join the free riders. Businesses that train for the good of the industry will be at a competitive disadvantage. As a result, the industry as a whole trains insufficient workers, and increasing numbers of businesses suffer shortages of skilled employees. And there is nothing any individual business can do about it if it is to remain competitive.

Free riding also undermines the ability of employees to band together to bargain with their employers for higher wages and better conditions.

Any improvements won by the bargaining will apply to all employees. So free-riding employees will benefit even though they do not lose wages in strike action, and do not risk retaliation from their employer. Again, the result is that the free riders win out, and cooperation is undermined.

Free riding occurs wherever individuals, whether they are molecular processes, cells, animals, humans or organisations of humans, can get the benefits of cooperation without contributing to its costs. Free riders will always end up ahead of cooperators who use energy or resources by cooperating. Wherever free riding is possible, it undermines cooperation.

Cells provide further examples. Cells that put their own success first will out-compete cooperative cells that sacrifice their own interests. Consider a number of cells that are teamed up to form a group. To take full advantage of the benefits of cooperation, the cells must develop a cooperative division of labour, with different cells specialising in different functions. However, once a cell significantly changes its structure and function to become an efficient specialist, it will no longer be able to reproduce to form a new organisation of cells. If a cell is to start a new group of cells, it must be able to differentiate to form all the types of cells in the new group. A cell that is already highly differentiated will not be able to do this. Only a cell that is non-specialised can start a new group and produce all the specialised cells.

So cells that help the group by specialising give up their chance of contributing to future generations. They will be unable to found new groups. As a result, they will not appear in the next generation, and will die out. They will be out-competed by less specialised cells that contribute less to the group but retain the chance of establishing new groups. No matter how beneficial to the group, specialisation and differentiation will not evolve because it is not in the interests of individual cells. The great American evolutionary theorist Leo Buss has argued convincingly that a strong division of labour was not possible within multicellular organisms until this barrier to cooperation was overcome[7].

Another example involves the mitochondria, the organelles inside modern cells that release and convert energy for the use of the cell. It is now widely accepted that mitochondria are the descendants of free-living bacteria that took up residence within cells[8]. To a certain extent, the first mitochondrial bacteria to live inside cells had interests in common with their hosts: the bacteria could do useful things the cell could not, so they could make the cell more efficient. The cell provided food and a protected environment for the bacteria. And if the bacteria helped the cell to do better, the bacteria did better as well.

But the very close cooperative relationship that exists between mitochondria and modern cells has not been achieved by either sacrificing their interests to the other. In fact, there is clear evidence that mitochondria which do not put their interests before those of the cell will be out-competed by mitochondria that do. And this is true even if the self-interest actually damages the interests of the cell[9].

Consider the following example: in most multicellular organisms, an individual obtains all its mitochondria through the egg cell from its mother[10]. The sperm cell from the father usually does not contain any mitochondria. So, from the point of view of mitochondria, the production of male offspring is a waste of time and resources. The mitochondria that a female passes to a son will not be passed to his offspring. They will die with the son. In contrast, the more daughters that a female produces, the more mitochondria it will pass to the next generation. If a female puts all its energy into producing daughters, its mitochondria will do better, and will increase in numbers in the population.

Of course, it is not in the best interests of the organism itself to produce only daughters. Its genes will generally do better if it produces an equal number of sons and daughters. But the interests of the organism has not stopped mitochondria in a number of species from manipulating their host organism to produce a higher proportion of females. For example, in some hermaphrodite plants, the genes of the mitochondria operate to inhibit the development by the plant of male pollen, and instead enhance the production of the female seeds that pass mitochondria to the next generation[11]. These mitochondria will out-compete any mitochondria that cooperate with the plant and do not pursue their own interests at the expense of the plant.

Self-interest is also the best strategy for molecular processes that reproduce and compete. We will consider the case of autocatalytic sets formed of protein molecules. Such sets are thought to be very important in the origin of life because they are probably one of the simplest ways in which protein molecules can get themselves replicated[12].

The basic members of an autocatalytic set are protein molecules that are able to catalyse each other's production. That is, they are able to speed up reactions that might not otherwise occur at a significant rate, and these reactions produce protein molecules that are also members of the set. Catalysts can do this by collecting together smaller molecules and holding them in positions in which they are likely to react. In this way they can organise reactions that would be extremely improbable if they occurred only by the chance meeting of the smaller molecules.

The set as a whole becomes autocatalytic and self-replicating when the formation of every member is catalysed by some other member of the set. Every member of the set gets reproduced by the catalytic activity of other members of the set. Once a set of protein catalysts links up in this way, it gains the ability to reproduce itself and its members indefinitely through time. This would be the case even though none of the members could survive indefinitely alone.

The particularly exciting thing about autocatalytic sets is that they form very easily. American biologist Stuart Kauffmann has shown that once you get a mixture of proteins that contain enough different types of proteins, it is likely that they will form a self-replicating autocatalytic set[13]. The chances are that the mixture will contain a set in which each component is catalysed by another component. These different proteins do not have to be specially selected. Even if they are chosen at random, if there are enough of them they are likely to include an autocatalytic set.

So once you get enough different proteins together, self-replication will arise easily. And the formation of these mixtures of protein should not be a rare and improbable event on a planet such as ours. On this basis, Kauffmann argues very persuasively that the fundamental step in the origin of life is easily taken, and is highly probable on this and other suitable planets.

Once a number of these sets form in a particular location, they will compete with each other for the resources needed to build the members of the set. The most competitive sets will win out. However, within the set itself, cooperative molecules that contribute most to the competitiveness of the set as a whole will not necessarily survive. They will be out-competed by molecules that are better at looking after their own interests, even where this damages the interests of the set as a whole[14].

For example, consider a cooperative catalyst that greatly improves the effectiveness of the set as a whole. It does this by catalysing the production of a molecule used in the construction of many other components of the set. But it will not survive in the set unless its own production is catalysed, no matter how useful it is to the set as a whole. And it will be out-competed by a similar molecule that uses its catalytic ability to further its own interests instead of the set's. Self-interested molecules do not catalyse molecules that are widely used in the set, but of no use to themselves. Instead they catalyse the production of molecules that in turn catalyse the reproduction of the self-interested molecules themselves.

We can conclude confidently that the barrier to the evolution of cooperation applies to all living processes. The circumstances that cause it are universal. Individuals who use resources to help others without benefit to themselves will be out-competed. They will be disadvantaged compared with those who use the resources for their own benefit. And the barrier applies no matter what the evolutionary mechanisms that adapt and evolve individuals. In the examples we have considered, the barrier has applied whether the evolutionary mechanisms are those that adapt corporations, individual humans, other multicellular organisms, single cells, or autocatalytic sets.

This barrier makes it difficult for evolutionary mechanisms to take advantage of the potential benefits of cooperation. It makes it difficult, but not impossible. If the barrier completely prevented the evolution of cooperation, evolution would not progress. But as we shall see in the next three Chapters, the barrier is not insurmountable. Evolution can exploit the advantages of cooperation by finding ways to make cooperation pay for the individuals who cooperate. To evolve, cooperation must be organised so that it is in the interests of individuals.

So if evolution is to progress, it must meet this central challenge: it must discover ways of building cooperative organisations out of self-interested components—it must learn how to make it in the interests of individuals to cooperate.

As we have seen, evolution has had great success at this already. It has discovered how to build complex cooperative organisations of molecular processes and highly cooperative organisations of cells. Evolution on earth has been progressively producing cooperative organisations of wider and wider scale.

But so far evolution has had only limited success in building cooperative organisations of greater scale than multicellular organisms. To date, human society is as far as it has progressed. However, as we have seen, human organisation is nowhere near as spectacularly cooperative as the components of cells and as the cells within animals. The potential for beneficial cooperation amongst humans and other organisms on this planet is far from exhausted, and will continue to drive progressive evolution. We are only a step along this way. We are evolutionary work-in-progress. And as we shall see in detail, evolution's next great challenge on earth is our challenge. To progress, evolution must discover ways to expand and improve cooperative organisation on this planet. Whether this occurs in the near future depends on us. The only evolutionary mechanisms on earth that are capable of discovering these improvements operate through our minds and our social systems.

In the next Chapter we will begin to look at how past evolution has found ways to overcome the barrier to the evolution of cooperation. We will see how evolution has repeatedly discovered ways to build cooperative organisations of living processes out of self-interested components. In later Chapters we will use this understanding to see how humanity can build larger-scale cooperative societies that can further exploit the advantages of cooperation to achieve future evolutionary success.

5

Organising Cooperation

Cooperation enables living processes to do better in evolutionary terms. Whatever evolutionary challenges they face, organisms will do better by cooperating. And the wider the scale of the cooperation, the more effective it is—the greater the power the cooperators will have over their environment, living and non-living. When evolution discovers ways to increase the scale of cooperation, life progresses in evolutionary terms.

But cooperation does not evolve easily. Self-interest stands in the way. Wherever cooperation has evolved, this impediment has had to be overcome. If we are to demonstrate that evolution can progress by organising cooperation over wider and wider scales, it is not enough to show only that wider-scale cooperation delivers evolutionary advantages. We must also show that evolution is capable of organising cooperation to exploit the advantages. How has evolution got around the barrier to the evolution of cooperation? How has it built cooperative organisations out of self-interested components? And can evolution continue to progress in this way in the future, building cooperative organisations of even larger scale?

Until recently, when evolutionists have attempted to answer these questions, they have focused on the obvious and simple instances of cooperation that we see around us. They have concentrated on the limited cooperation we see amongst multicellular animals such as mammals, birds and insects. It is only in the last decade or so that much work has been done on the evolution of the more complex cooperation found within cells, within multicellular organisms, and within human society[1].

Evolutionary theorists have been very successful at explaining the limited cooperation that exists amongst some multicellular organisms. Theorists have shown that there are a number of mechanisms that will enable simple cooperation to evolve amongst these animals. And they can explain why the mechanisms have produced cooperation only amongst a relatively few species. Current evolutionary theory is good at showing why most of the animals we see around us do not cooperate much.

As might be expected, the mechanisms are not so good for explaining the evolution of the spectacular cooperation we see within cells, multicellular organisms, and human society. The mechanisms do not disclose a general method for organising cooperation amongst self-interested components that would work for any living processes in any circumstances. The mechanisms are therefore not much help in showing how to build human organisations that are more cooperative.

However, the discovery of these mechanisms has been an important step towards a comprehensive theory of how evolution can overcome the barrier to cooperation. While the mechanisms are not a comprehensive answer, they point to the direction in which a general answer might lie.

The first major breakthrough in understanding how cooperation might evolve between multicellular animals came with the publication of William Hamilton's theory of genetical kin selection in 1964[2]. The central idea in the mechanism he described is very simple: consider a cooperator who shares its food with others who have not been able to gather enough. If the cooperator shares food only with other individuals who contain copies of its cooperator gene, the benefits of cooperation will go only to other cooperators. Only those who carry the cooperator gene will benefit. Non-cooperators will be excluded from any of the advantages of cooperation.

Provided the benefits of cooperation outweigh the costs, cooperators will end up in front, and will out-compete non-cooperators. Even if an individual cooperator ends up worse off because it pays the costs of cooperation and gets no benefits, cooperator genes will be better off overall—a copy of the gene will also be carried by the individuals who get the benefit of cooperation. The full benefits of cooperation will be captured by the cooperators, the barrier to the evolution of cooperation will have been overcome, and the potential advantages of cooperation can be exploited.

Of course, the catch is that individuals cannot tell easily which other individuals carry the cooperator gene. So they are unable to direct their cooperation only towards others who carry the cooperator genes. Some of the benefits will leak to free riders and other non-cooperators, enabling them to do better than cooperators.

Hamilton saw that there was a partial way around this. If an individual carries the cooperator gene, its relatives are more likely to carry the gene. So if you are a cooperator, and you also know who your relatives are, you know individuals who are more likely to be cooperators. And you do not even have to know exactly who your relatives might be. If you are an organism that tends to live its life near where it is born, you can be sure that those around you are more likely to be your close relatives.

But the less certain you are about who else carries a copy of your cooperator gene, the less effective is kin selection at evolving cooperation. For example, if you carry a cooperator gene, there is only a 50 per cent chance that your brother or sister also has the same gene due to relatedness. So if you provide some resources to a brother or sister, you are denying your cooperator gene the value of those resources, in exchange for only a 50 per cent chance that the resources will be going to a related cooperator gene. If the cooperator genes are to end up in front, you would help a brother or sister only if the benefits of the cooperation outweigh the costs by at least two to one. When you help a brother or a sister, there is a fifty-fifty chance you will not be helping a related cooperator gene. So on the 50 per cent of occasions when the brother or sister does carry the cooperator gene, you have to get double the return. You have got to make up for the 50 per cent of times when the cost of cooperating is wasted.

To take a starker example, consider a cooperator who sacrifices its life to help its brothers and sisters. On average, it would have to save at least two of their lives to be sure that it was saving at least one related cooperator gene to replace its own.

The more distantly-related the individuals, the worse this gets. The pay off from cooperation has to be even higher for cooperation to come out in front. For example, it is worth helping a first cousin only if the benefits produced by the cooperation outweigh the costs by at least eight times. On average, there would be one related cooperator gene in eight first cousins. An individual cooperator should sacrifice itself for a first cousin only if it saves at least eight of them.

As a result, kin selection is a very poor mechanism for evolving cooperation. Because relatedness is an imperfect indicator of common genes, kin selection can evolve cooperation only where the benefits from cooperation far outweigh the costs. This leaves an immense amount of beneficial cooperation that cannot be evolved by kin selection. If a mechanism is to be able to fully explore the benefits of cooperation, it must be able to establish any cooperation in which the benefits outweigh the costs. Kin selection falls far short of this. Without these severe limitations, kin selection would have been much better at organising cooperation amongst the animals around us. The potential benefits of cooperation would have driven the evolution of a complex division of labour within each population of animals, similar to what we see within our societies, our bodies, and our cells.

But as a theory, kin selection has been very successful. The ineffectiveness of the kin selection mechanism is consistent with the sparse and limited cooperation found amongst multicellular animals. And where there is cooperation, the way it is organised is consistent with kin selection. Much of the cooperative organisation we find amongst the animals around us is between relatives[3].

The most common is where parents provide food and protection to their young. But parents are not the only individuals that help to bring up young in some species. In a number of bird species, eggs are incubated and young are fed by birds that are often not their parents. An example that has been studied at length is the Mexican Jay that lives in flocks of five to fifteen in pine and oak woodlands. Each flock defends a territory as a group, actively excluding other flocks. At each nest most or all of the members of the flock help bring food to the young. Birds that are not parents typically bring about half of the food[4].

Why do these birds cooperate in this way? Surely they would be better off in evolutionary terms by eating the food themselves, and putting their time and energy into starting up their own families? Kin selection provides the answer to this. The individuals within each flock are highly related, and the cooperators are in fact helping others that have a higher likelihood of carrying their cooperator genes.

Similar patterns are found in other birds that breed communally, including Australia's superb blue wrens, and white winged choughs[5]. High relatedness is also a universal within the cooperative hunting groups of African lions, wild dogs, and spotted hyenas, and within the groups of savannah baboons and many species of monkeys that collectively defend against predators[6]. And the highly cooperative colonies of ants, termites, and bees are invariably formed of individuals that are closely related[7].

The second mechanism that enables cooperation to evolve between animals is reciprocal altruism[8]. Again, the basic idea is very simple: an individual will not be disadvantaged if it helps another provided the other helps it in return. The individual's initial costs of cooperating will be recouped when the other returns the favour. If the benefits of cooperation outweigh the costs, both individuals can end up in front. The cooperators will therefore out-compete individuals that don't reciprocally help each other in this way.

For example, when an individual is successful in collecting food, it might share it with others who have not been so lucky. On occasions when others are more successful, they can return the previous favour by sharing their food. Over time, all will be better off than if they did not share.

Exactly this sort of cooperation has been found in South American vampire bats[9]. Each night the bats leave the hollow trees where they spend the day to search for large animals. If a bat finds an animal, it will attempt to feed on the animal's blood. But there is no guarantee that an individual bat will successfully find and feed off an animal every night. Young inexperienced bats are unsuccessful about one night in three. If they have a few bad nights in a row, they can be in danger of starving. This danger will be avoided if the bats that hole up together share blood. Those that have been successful regurgitate some blood for others that have missed out. Because all bats risk having nights on which they fail to get blood themselves, they can all benefit from this cooperation.

But closer study of the bats showed that any particular individual would not regurgitate blood for all of the other individuals. It would give blood to some, but not to others. And those who got the blood were not necessarily close relations, as would be expected if kin selection were operating. It turned out that those who were given blood were bats who had given the individual blood in the past. And those who were refused blood were those who had previously refused the individual blood. The bats knew each other as individuals, knew who were likely to return favours and who were not, and chose to give blood only to those who were likely to reciprocate.

A closer understanding of reciprocal altruism and of its limitations shows that the bats have to be this choosey. In fact, reciprocal altruism cannot be established in a group of organisms unless they are capable of refusing to cooperate with individuals that fail to return favours. If an individual continues to do favours for others that do not return the favour, the non-reciprocators will win out. An individual that cheats by receiving favours but not repaying them will gain all the benefits of cooperation without paying any of the costs. And if a cheat is within a large population of cooperators, it will do extremely well, collecting favours and never returning them. In these circumstances, cheats will prosper, increasing in numbers in the population until all cooperators die out. In such a group, the most competitive strategy will be to accept favours but not reciprocate. Reciprocal altruism will not survive, even if its benefits far outweigh its costs.

However, all this changes if the individuals live together for a long time, repeatedly have the opportunity to exchange favours, and have the capacity to remember which individuals return favours, and which do not. If these conditions are met, cheats will not be able to continually take benefits from the cooperators, because they will be recognised as cheats and excluded from further cooperation. Once this happens, they will not gain any of the benefits of cooperation. In contrast, cooperators will end up in front because they will limit their exchange of favours to those who have shown they will return the favours. If they cooperate in this way whenever the benefits outweigh the costs, cooperators will out-compete cheats[10].

But it will rarely be this simple. It might pay cheats to behave like cooperators for a while to extract favours, then to quit while they are in front. Or when they have gained a reputation for cheating in one group, they may move on to another group where they are unknown. To avoid being cheated, cooperators will have to get very good at detecting cheats[11].

Furthermore, the whole system of reciprocity works only if there is a fair balance between the favours that are exchanged. Individuals whose favours cost more than the individuals receive in return will be out-competed. But most animals have little ability to judge whether reciprocal favours are similar in value. And having to do so further complicates the critically important job of telling the difference between cheats and cooperators.

All these challenges become almost impossible when acts of cooperation benefit many others, and impact differently on each of them. If a cooperative act benefits a number of others in the group, rather than a specific individual, it is much harder to work out who owes who what, and how much. But unless individuals can keep track of all this, cheats will flourish

The great advantage of reciprocal altruism over kin selection is that it can evolve cooperation between individuals who are not related. However, the conditions under which it works are so limited that it has not been very significant in establishing cooperation between animals. Not many animals live together for long periods, repeatedly have the opportunity to cooperate, are able to recognise each other as individuals, can remember how each individual behaved each time it had the opportunity to cooperate, and can keep track of favours owed and favours earned. Cases of reciprocal altruism have been found only in a few species of animals such as vampire bats, dolphins, and elephants, and in some species of monkeys and apes[12].

Of course, the species that meets these conditions to the highest degree is us. And the theory of reciprocal altruism fits very well with the way we relate and cooperate with people that we meet repeatedly.

It works well for explaining key aspects of the cooperative relationships between people in tribal groups, closely knit communities, and in friendship networks and other stable social groupings.

The theory predicts that successful reciprocal altruists will have two great concerns in their relationships with others. First, they will not want to be cheated by others, so they will want to know who can be trusted to return favours. Second, they will want others to choose them as cooperative partners, so that they can share in the benefits of cooperation. They will want others to see them as trustworthy cooperators. They will not want to be seen as a cheat that should be excluded from beneficial cooperation[13].

Consistent with this, we humans have many behaviours that seem to serve the function of protecting us against cheating. For example, we tend to have an extraordinary interest in judging other people's trustworthiness and honesty. A major part of gossip and other social communication involves swapping information that helps us to make and update these judgements. We are also reluctant to risk cooperating with strangers when we have no knowledge of their reputation and character. We are even more reluctant when we know the stranger does not depend on us for future cooperative opportunities. In these circumstances, they will lose little if they destroy their reputation with us.

We are also very concerned to protect our own reputation for trustworthiness and honesty, particularly amongst those that we spend a lot of time with. We often react instantly and emotionally to any attack on these aspects of our character. Most of us are also concerned to ensure that we are seen to treat our friends fairly whenever there are things to be shared, and that we repay our obligations to people we know. And we do not want to enter into obligations with friends that we might not be able to repay.

But even amongst humans, the capacity of reciprocal altruism to establish cooperation is severely limited, even where cooperation produces substantial benefits. As we have seen, reciprocal altruism will work only where cheats can be successfully excluded from future cooperation. This will not be the case if cheats can easily find other individuals to deal with who know nothing of their previous actions. It is only where individuals have no option but to continue living and interacting together for long periods that cheating will not pay. If an individual earns a reputation for cheating in these circumstances, it will not find other cooperators to deal with.

This condition is not generally met in modern large-scale economic markets. Individuals can readily leave a bad reputation behind them, and find new individuals to deal with who are unaware of their previous actions. Reciprocal altruism is therefore not the mechanism that has produced the large-scale markets that organise economic exchanges across human societies.

The other great limitation of reciprocal altruism is that it is only effective where the benefits of cooperation can be neatly parcelled up, precisely valued, and swapped between individuals. Where the benefits of a cooperative act impact on many individuals, reciprocal altruism cannot operate easily[14]. This is a significant limitation. In the complex, differentiated cooperative organisation that are cells, organisms, and human societies, most cooperative acts impact on many others in the organisation. It would be impossible to separate out the effects and keep track of them as favours that have to be returned. And if favours are not returned with acts of similar value, reciprocal altruism fails.

So genetical kin selection and reciprocal altruism are not the mechanisms that have enabled evolution to explore the potential benefits of cooperation within cells, multicellular organisms, and human society. And they are not the mechanisms that will change humans from partial cooperators into full cooperators. They can achieve only limited cooperation under restricted conditions. If kin selection and reciprocal altruism were the only mechanisms that evolution could use to organise cooperation, evolution would not be progressive. It would not be able to exploit the benefits of cooperation by progressively building cooperative organisations of wider and wider scale.

But kin selection and reciprocal altruism do hint at a more general way of building cooperative organisation out of self-interested components. Where they are effective, both mechanisms establish cooperation in the same way. They ensure that cooperators benefit from their cooperation. The mechanisms work to the extent that cooperators are able to capture the benefits created by their cooperation[15]. Kin selection achieves this to the extent that the individual helped by the cooperator also carries the cooperator gene. If it does, the cooperator gene captures the benefits of the cooperation. Particular cooperators may not capture the benefits of their cooperation, but cooperators as a class will. If the cooperation is worthwhile, its benefits will outweigh the costs, and the cooperator genes end up in front.

Reciprocal altruism works to the extent that the recipients of cooperation return the favours. If they do, cooperators capture the benefits of their cooperation. They benefit from all of their cooperative acts, because each will earn a return favour. If the cooperation is worthwhile, cooperators again end up in front.

Where the mechanisms fail, both do so for the same reason. They fail to the extent that cooperators do not capture the benefits of their cooperation. Kin selection is undermined when the individual helped by the cooperation does not carry the cooperator gene. When this occurs, the cooperator gene will not capture the benefits of cooperation, only the costs, and will be out-competed. Reciprocal altruism is undermined when a recipient of cooperation cheats by failing to return the favour. When this occurs, a cooperator will not capture the benefits of its cooperative act, only the costs, and will end up worse off than the cheat.

This analysis of kin selection and reciprocal altruism points to what a mechanism must do if it is to be capable of establishing cooperation whenever cooperation is worthwhile. It must ensure that cooperators capture the benefits of their cooperative actions. Free riders, cheats and thieves who contribute nothing to the cooperation must capture none of the benefits created by cooperation.

To overcome the barrier to cooperation fully, a mechanism must also ensure that those who harm others capture the effects of any harm they cause. Harming others must be harmful to those who cause it. And to organise cooperation across generations, a mechanism must also ensure that all the effects of actions on others must be captured, no matter how distant in the future the effects may be. This is necessary to ensure that cooperation that benefits future generations will also be profitable to those who produce it.

A mechanism that ensures individuals capture all the effects of their actions would completely overcome the barrier to the evolution of cooperation. If individuals capture the effects of their actions on others, self-interest would no longer stop an individual from helping another. Helping the other would be as profitable to the individual as helping itself. As a result, the individual would treat the other as it would treat itself. And it would do so without giving up its self-interest in any way. It would be in the interests of the individual to treat the other as self[16]. Anything it could do to help the other would help itself. Anything it could do to harm the other would harm itself. The individual would benefit as much from discovering new ways to help others as it would from finding new ways to help itself directly. All this holds true whether it is individual cooperators or cooperators as a class who capture the effects on others of their actions.

At this point it is useful to look briefly at the difference such a mechanism would make for the examples of failed cooperation that we looked at in the previous Chapter. Taking each example in turn, it would ensure that: corporations that pollute would capture the harm they visited on others, causing them to be out-competed by corporations that introduced less efficient but less harmful manufacturing processes; businesses that trained employees would capture the benefits of producing skilled employees for their industry; only employees who contributed to the bargaining process with their employer would obtain any improvements in wages and conditions that were won through the bargaining; cells that specialised to perform useful functions for the group of cells would out-compete cells that contributed less to the group; mitochondria that harmed their host would also harm their own interests, and would be competitively disadvantaged against other mitochondria; and protein enzymes that contributed more to the autocatalytic set as a whole would in turn have their production catalysed by the set.

Wherever complex cooperation has been able to evolve, it is because cooperators have been able to capture the effects on others of their actions. As a result, self-interest has driven them to treat the other as self. A strong example that may not be immediately obvious is an economic market. The market will reward an individual who develops a new product that benefits others by satisfying their needs better than existing products. Selling the product enables the individual to capture the beneficial effects that the product has on others. In the limited areas where economic markets work effectively, individuals benefit in this way from actions that benefit others. Where the market enables individuals to capture the benefits of their effects on others, the market aligns the interests of individuals with the interests of others.

In contrast, where the barrier to the evolution of cooperation has not yet been overcome, cooperators do not capture the beneficial effects on others of their actions. For example, in our economic system, a person will not capture the benefits of providing food to people who are starving in an African famine. It would be far more profitable to invent an improved mousetrap. Or a new weapon of mass destruction. But when the barriers to cooperation in human society are overcome, and when all individuals capture in full their beneficial effects on others, the provision of food to the starving will be a lucrative way to make a living.

So a mechanism that ensures that cooperators capture the effects of their actions on others would allow cooperation to be established wherever cooperation is worthwhile. Organisms that pursue only their own interests would act cooperatively. The mechanism would establish cooperation between organisms whose basic nature is greedy and selfish. It could produce cooperation between lawyers and real estate salesmen. The mechanism would enable the potential benefits of cooperation to be fully explored and exploited amongst any living processes. It would enable evolution to progressively exploit the advantages of cooperation, organising cooperative organisations of wider and wider scale.

This is the ideal. But is there a mechanism that can achieve this? Is there a general form of organisation in which individuals capture the effects on others of their actions? Or is the ideal unattainable?

One general way in which cooperators could capture the effects of their actions on others is if another individual takes action to ensure this happens. The other individual could do this by providing cooperators with benefits that reflect the useful effects of the cooperators on others, and by punishing those who harm others.

To see how this might work, imagine a population of organisms that are formed into a number of unorganised groups. Within each group, the barrier to cooperation would apply. Individuals who use resources to help others without benefit to themselves would be less competitive than non-cooperators. A group that contains cooperators might do better than other groups for a while, but this would not help the cooperators. Within such a group, the cooperators would still be less competitive, and they would still die out. Specialisation and division of labour could not evolve within a group no matter how beneficial this would be to the group as a whole.

What would be the outcome if a group contained an individual who acted to ensure that all others in the group captured the effects of their actions on others? The individual would provide resources and services to cooperators to reflect the benefits of their cooperation. And the individual would punish free riders, cheats and thieves who would otherwise take the benefits of cooperation without contributing anything to others in the group.

Within such a group, cooperators who produced net benefits for others would end up in front. They would out-compete free riders, cheats and lesser cooperators. Whenever a more effective form of cooperation was discovered, it would flourish within the group. Specialisation and division of labour would be established wherever it was beneficial.

But there is a fundamental flaw in this.

An individual who was a normal and typical member of the group would be disadvantaged if it carried out this task of rewarding cooperators and punishing cheats. An individual who used its time and resources in this way would be out-competed by those who instead used their resources for their own benefit. Although the individual would be carrying out a function that would greatly benefit the group as a whole, it would disadvantage itself in the process. It would run into exactly the same barrier as any other individual who does things for the benefit of the group. And there would be no other individual to reward it so that it captured the beneficial effects of its actions on others[17].

The problems do not end there. As a typical member of the group, the individual would also be unable to gather all the resources needed to reward cooperators for their contributions. When it attempted to collect resources for this purpose the individual would be in competition with all other members of the group. It would not be able to gather more than others.

And the individual would not have the capacity to control and punish cheats and free riders in the group. It would not be in the evolutionary interests of cheats to be controlled and punished. They would have a very strong incentive to escape control, and to exploit the growing cooperation in the group. A typical member of the group would not be able to control cheats, free riders and thieves, particularly if they ganged-up to resist control.

A typical member of the group cannot carry out the function of organising the group so that members capture their effects on others. How could this difficulty be overcome? What sort of entity could successfully organise the group in this way, and enable the immense benefits of cooperation to be exploited?

If the job of organising a group of humans in this way was on offer, no one would be likely to accept the job unless they also had the power and authority needed to do it. How they got the power would not be of critical importance. The job would probably be easier if they got the authority by the agreement of the members of the group. But the job could also be done even if the authority was achieved by coercion, without agreement.

This points to a general way in which these problems could be overcome, enabling the group to be organised to promote cooperation. The entity or team of entities who organise the group must have control over the group. It must have the power to control and manage the individuals who make up the group. This controller, who I shall call the manager, must have the capacity to take from the group whatever resources it needs to carry out its tasks. And it must be able to punish others without risk to itself[18].

Importantly, a manager who has this power would not be in competition as an equal with the other members of the group. Other individuals in the group would not have the capacity to out-compete the manager because the manager has control over them and their resources.

With this power, the manager has the ability to take whatever resources it needs to give to cooperators to ensure that they capture the benefits of their effects on others. The manager will be able to redistribute resources within the group to ensure that cooperation pays. And the manager will have the power to control and punish cheats and free riders throughout the group, even when they gang up against the manager. It can overcome all the problems that stop a normal member of the group from organising the group to support cooperation.

I will now develop a number of broad examples that will illustrate more concretely how this form of organisation is able to support cooperation. I will begin with an example in which cooperation is sustained within human organisation by the actions of a manager. Then I will look at how the management of molecular processes and mitochondria has been able to produce the complex cooperation found within cells.

About ten thousand years ago, humans began to domesticate plants and animals and form the first agricultural communities. Potentially, the communities could benefit greatly from specialisation, division of labour, and other forms of cooperative activity. For example, it would be more efficient if some individuals specialised in making farm implements or weapons, if others trained animals for use in transport and farming, and if others specialised in defending the community. And the welfare of the community would be greatly enhanced if individuals cooperated together to construct irrigation systems for farming, or to fight off enemies.

But the establishment of these and other forms of cooperation had to overcome the barrier to the evolution of cooperation. If the barrier was not overcome, cheating and stealing would make it difficult for specialists to get a proper return for what they produced. And free riders could avoid making contributions to community activities such as defence and irrigation, but could still share in the benefits produced by community projects.

As a result, cooperators would not capture all the benefits of their cooperation. And it would not be in the interests of any individuals in the community to initiate many forms of cooperation that could produce significant benefits to the community. The community could not exploit the immense benefits of cooperation. However, this could be completely turned around if the community was managed by a powerful ruler. The ruler could collect taxes off all members of the community, and redistribute these to support useful cooperation. The ruler could pay for specialist services and fund other activities that the ruler judged were of benefit to the community as a whole. And the ruler could punish thieves and cheats[19].

A particularly wise ruler might discover a better alternative than using tax to support specialists such as toolmakers. The ruler might find that specialists would be able to make a living by exchanging their products for food and other services, provided they were protected from theft and from cheats who do not reciprocate in exchanges. The ruler could provide this protection. Cheating and theft could be punished, exchange agreements could be enforced, and disputes could be heard by the ruler and settled fairly. With these protections, a system of market exchange would emerge.

Once a market system existed, the ruler would not have to decide the types of goods and services that were to be produced by specialists, nor their quantities and prices. Instead, these issues would be decided directly by producers and consumers through their interactions. The advantage is that producers and consumers would be in a much better position than the ruler to make these decisions. The consumers would be far better able to assess the value to them of what the specialists produced. And the specialists would be far better placed to decide whether it was worthwhile for them to produce particular goods and services, given the costs of doing so and the price they could get.

However, such an exchange system could not work effectively without a ruler who could manage cheating and stealing. Without this, all that could emerge is a system of reciprocal altruism, with all the flaws and limitations of such a system. The essential role of the manager is to patch up the limitations in the mechanism of reciprocal altruism so that cooperators are able to capture the benefits of their acts. And the individuals involved in the exchanges are unable to do this patching. Only managers who have power and control over individuals can do it. As we shall see in detail later, a manager is as essential to a market economy as it is to a centrally planned economy[20].

In these ways, a manager can produce a highly cooperative organisation of humans, even if the individual members of the organisation are purely self-interested.

The barrier to the evolution of cooperation in autocatalytic sets can also be overcome by management. As we have seen, a protein enzyme that catalyses reactions that greatly improve the efficiency of an autocatalytic set would not necessarily be catalysed in return by other members of the set. In fact, such a cooperative protein could be out-competed by a similar protein that did not help the set as a whole. This would be the case if the non-cooperator protein used its time and resources to catalyse other proteins that in turn catalysed the production of the non-cooperator.

The management of a set by one or more RNA molecules could change this. RNA molecules could take control of an autocatalytic set of proteins because of their larger size, greater stability, and their superior ability to catalyse a diversity of proteins. The RNA could use its superior catalytic ability to cause the formation of proteins that otherwise would not exist in the set. It could use this ability to produce cooperative proteins that are beneficial to the set as a whole. A specialist protein could be supported by the RNA to the extent needed to reflect the beneficial effects of the protein on others in the set.

The RNA could also produce proteins that further enhanced the ability of the RNA to catalyse other proteins, improving its ability to control the set. And it could control cheats and free riders that take resources from the set and are catalysed by the set without contributing anything in return. The RNA could do this by catalysing reactions that take resources and catalysis away from the cheats and free riders. The RNA could have as much control over what happens in an autocatalytic set as a human ruler who controls an agricultural community.

The RNA manager could use its control over the set to ensure that each type of protein captures the effects of its actions on the set as a whole. The manager could ensure that its support for the production of a particular type of protein reflects the contribution of the protein to the organisation.

Management was also essential for the evolution of the close cooperative relationship that developed between cells and the bacterial ancestors of mitochondria that began to live inside the cells. As we saw in the previous Chapter, bacteria that put all their time and effort into helping the cell in which they lived were not likely to have been successful. They would have had to compete with other bacteria who instead plundered the resources of the cell, used them to breed up, then left the cell to find other cells to plunder. These non-cooperative bacteria would free ride on any help given to the cell by cooperative bacteria.

But cells could overcome this by controlling the bacteria inside them. A critical step was for cells to prevent the bacteria from leaving them[21]. Once this option was cut off, the fate of the bacteria was much more closely linked to the fate of the cell. Free riders that were a drain on the cell would now capture their harmful effects because their future was totally dependent on the success of the cell and its descendants. Free riders would lose out. Cooperators could also now capture the benefits of their cooperative effects on the cell. The better the cell and its descendants went, the better the cooperators and their descendants went.

Multicellular organisms had to introduce additional controls to prevent mitochondria competing within them and undermining cooperation. A key control was to prevent the transmission of mitochondria from males to their offspring[22]. As we saw earlier, mitochondria are transmitted only through the egg, not through sperm. So all the mitochondria in the cells of a multicellular organism such as yourself are descendants of mitochondria that came from the mother. And the mother's mitochondria are all descendants of those that came from her mother, and so on. There is no mixing of mitochondria from different individuals or different lineages.

This is important for ensuring that cooperative mitochondria win out: if mitochondria are prevented from mixing with mitochondria from other lineages, they do not have the opportunity to free ride off them. If males did transmit mitochondria, and mitochondria from different lineages did mix, free riders could flourish and undermine cooperation. They could free ride off any cooperative mitochondria they were mixed with, out-competing them in the process. Then they could move on when the host reproduced, to be mixed with a fresh lot of cooperators, and so on. The result would undermine cooperation amongst mitochondria, and cooperation between mitochondria and the cells that house them.

There is clear evidence that different lineages of mitochondria would compete and undermine cooperation in this way if they were mixed: often, not all mitochondria are excluded from some sperm cells. A few include a small number of mitochondria. When such a sperm fuses with an egg cell, the mitochondria and other organelles in the sperm are usually destroyed immediately by their more numerous counterparts within the egg[23].

But locking mitochondria into separate lineages so that they eventually capture their effects on their hosts is not enough. Mitochondria within an individual host will still compete amongst each other to get into the egg cells that produce the host's offspring. And, as ever, mitochondria that put all their time and energy into helping the host's cells will be out-competed. Mitochondria that instead put their time and energy solely into getting into egg cells will win out.

A host organism will do better in evolutionary terms if it is able to manage this competition by supporting cooperators and preventing non-cooperators from gaining any competitive advantage. The host can use its power over its mitochondria to ensure that cooperators capture the benefits of their effects on the host, and that non-cooperators capture the harm they cause. The relationships that now exist between mitochondria and the cells they inhabit seem harmonious, consensual and mutual. But a closer examination shows that this has been achieved only by the strict control of mitochondria by the cells and organisms that house them. They can use controls to support cooperators, suppress cheats and free riders, reduce the possibility that non-cooperators will be able to get any advantage over cooperators, and even to restrict the likelihood that non-cooperators will arise by mutation.

Specific examples of how mitochondria are controlled by the cell that houses them include[24]: the rate at which mitochondria release energy for the cell is controlled not by the mitochondria, but by factors outside them in the cell; proteins produced by mitochondria cannot function independently—they work only in the presence of proteins produced by the cell; a large part of the genetic material that was contained in the bacterial ancestors of mitochondria no longer remains in the mitochondria—it is now contained in and controlled by the cell nucleus, leaving mitochondria with little control over their own functions and, more importantly, with little ability to change them; and finally, the genetic material that controls the operation of mitochondrial genes and their replication is no longer in the mitochondria—the genes of the cell control these critically important functions.

Cells have had to strictly control and manage mitochondria to ensure that the only way they can pursue their own interests is by helping the cell. To make mitochondria manageable, their independence has been greatly restricted, and their capacity to independently innovate and evolve has been suppressed by the removal of much of their genetic material. So that they serve the interests of their hosts, mitochondria have been enslaved and lobotomised.

These three examples illustrate how a manager can control a group of individuals in ways that enable the potential benefits of cooperation to be explored. Left to themselves, a group of individuals that interact as equals is unable to establish complex cooperative organisation. It will not be in the interests of any member of the group to initiate this cooperation, no matter how great the benefits that it can provide. This is because individuals will not capture the benefits of cooperation that they initiate, particularly in the face of cheating, free riding and theft.

This all changes when an extra layer of organisation is added to the group. This extra layer is the manager, made up of one or more entities who are able to control the group. The addition of this extra layer converts the original organisation of equals into what can be called a vertical organisation. Extending this terminology, the original organisation that had only one layer of organisation can be referred to as a horizontal organisation.

The manager can use its power to organise the horizontal group in whatever way is necessary to ensure that cooperators, free riders and cheats capture the effects of their actions on others in the organisation. To the extent that the manager is successful in doing this, the interests of all individuals will be aligned with the interests of the group, and individuals will treat others as they treat themselves. Self-interest will be the same as group interest, and the vertical organisation will have successfully created cooperative organisation out of self-interested components.

The manager of an ideal vertical organisation will be able to support and sustain whatever cooperative arrangements are best for the group. By determining which individuals succeed in the group and which don't, an ideal manager will be able to construct whatever organisation is needed for evolutionary success. The ideal manager will be able to make it in the evolutionary interests of members of the group to undertake whatever forms of cooperation are best. There will be no form of cooperative organisation that the manager cannot construct. Management will be able to produce any pattern of specialisation and any form of division of labour amongst the members of the group[25]. The barrier to the evolution of cooperation will be completely overcome.

It is worth emphasising that the manager must have the capacity to control the group if the manager is to be able to organise complex cooperative organisation. If the manager is to stop cheats, free riders and thieves, it must have power over them. It must be able to make them do things that would not otherwise be in their interests, and they must not be able to avoid this control by influencing the manager in return. If they can, they can escape control, and continue to undermine cooperation. This is what happens when criminals use bribery to influence police, and when friendships with staff undermine the ability of an executive to manage staff in a corporation.

Individuals that are equals within a horizontal organisation obviously cannot impose the level of control needed to stop cheats, free riders and thieves. Individuals that band together to act as a group can achieve some control over cheats. But when they do this they are no longer acting as individuals. They are forming a manager and a new level of organisation that can control individuals.

On closer examination, we find that even the limited forms of cooperation found between multicellular organisms cannot arise unless cheats and free riders are restrained somehow. If cheats and free riders are free to take advantage of cooperators, cooperation will be undermined. In the situations where cooperation occurs amongst multicellular organisms, we invariably find that there is something about the structure of the situation that restrains cheating and free riding. For example, in vampire bats cooperation would fail if cheats and free riders could move from hollow tree to hollow tree to continually exploit fresh cooperators who are unaware of their reputation. They are prevented from doing this because the bats within each hollow tree exclude other bats from using their tree. Bats need to defend their resting-places because good ones are in short supply, and other bats will try to take them over[26].

Similarly, kin selection would be undermined in birds that breed communally if cheats and free riders could easily join a group and take advantage of the cooperation. But, communal birds usually defend territories to obtain exclusive access to food and other resources[27]. So cheats and free riders are prevented from easily moving between groups, and closely-related birds will remain together. And finally, computer simulations show that reciprocal altruism can evolve easily amongst colonial animals that live fixed to rocks. With no mobility, cheats and free riders are prevented from moving freely from cooperator to cooperator, exploiting them as they go[28].

In all these cases, the factors that control cheats and free riders have not evolved because they help cooperation get established. They have evolved for other reasons, and the restrictions they place on the opportunities for cheating and free riding are chance side effects. So the restrictions are not tuned by evolution to be as effective as possible at promoting cooperation, and they will not adapt to stay effective when things change.

What vertical organisation does is take the control of cheats out of the hands of restrictions that have evolved for other reasons. As we shall see in the next Chapter, the controls established by managers are tuned by evolution specifically to promote cooperation. Controls that are best at promoting cooperation will win evolutionary struggles. And evolution will adapt the controls to find better ways to support cooperation and to prevent cheats and free riders from escaping control as they evolve.

For all these reasons, only vertical organisation has the capacity to support the extraordinary level of cooperation we find within cells, multicellular organisms, and human society.

But all we have shown to this point is that management has the potential to promote cooperation amongst others. Will it take up this potential? Will evolution establish managers that use their power to organise cooperation in the way I have described? Or will evolution favour managers that use their powers to plunder resources and services from other organisms, trampling over their interests in the process? Why would they use their power to promote cooperation between organisms? These are important issues for the case I am building in favour of progressive evolution. If it is not in the evolutionary interests of managers to organise cooperation, the barrier to cooperation cannot be overcome. And unless evolution is able to find comprehensive ways to overcome the barrier, it will not be able to progressively build cooperative organisations of wider and wider scales.

To a person at the turn of the 20th century, the answer to these questions might seem obvious. There is plenty of evidence from recent human history to show that if you give a person power over others, the power is likely to be abused. Power corrupts, and absolute power corrupts absolutely. In the 20th century individuals such as Stalin and Hitler who have had the most power seem to have been responsible for the most human deaths and misery.

Time and time again we have seen rulers and governments sacrifice the interests and often the lives of their people in order to stay in power. Many governments have been guilty of persecuting minorities on the basis of their religious beliefs, ideology or race. White South Africans and Australians are notorious for their exploitation and abuse of the indigenous peoples of their countries. And those who have the wealth to influence governments have never been shy about getting governments to use their immense power to favour them over others.

And there is abuse of power all the way down, at all scales. An extraordinary number of catholic priests have used their position to sexually abuse children, and, for most of the 20th century, the hierarchy of their church used its power to hide these practices and to protect the guilty. Even within families, individuals with greater power will often abuse others physically, sexually, emotionally or economically. And anyone who has worked in a large corporation knows that executives use their power over employees for many purposes that have little to do with the efficiency of the organisation.

These abuses have obscured the fact that the extraordinary level of cooperative organisation found within cells, organisms and human society has been achieved only through vertical organisation. But does this level of abuse mean that the ability of vertical organisation to establish cooperation is fundamentally limited? Is this abuse the price we have to pay for the benefits of cooperation that can be produced only by vertical organisation? Or is it simply that current human organisation and its obvious flaws should be seen as part of an evolutionary progression in which the flaws will eventually be overcome? Is this just an evolutionary stage we are going through? And if so, how will the flaws be overcome?

These are questions I will begin to answer in the next Chapter. We will see that vertical organisations of molecular processes and organisations of cells both passed through a phase in their evolution in which managers exploited the organisation, rather than promoted cooperation. But these phases ended when organisations emerged that were controlled by management that has discovered how to support cooperation. Management that used its power to organise cooperation proved superior in strictly evolutionary terms to management that used its power to exploit the organisation. We will see that competition between organisations aligns the evolutionary interests of the manager with the organisation. Managers that promote cooperation do better than those that do not, because the organisations they manage are more competitive.

But competition between human societies is no longer strong enough or incessant enough to align the interests of human rulers and governments with those of their citizens. When human groups were small and numerous, competition was strong and continual. But as human societies grew larger and fewer, competition between them has decreased. We will see that in the absence of strong and continual competition, the key to the evolution of cooperative human societies is the discovery of new organisational structures. These must align the interests of the ruler or government with those of the members of the society. When this is achieved, a government driven purely by self-interest will manage in the interests of the society. Abuse and exploitation will end, and the benefits of cooperation can be fully exploited. In the later Chapters of this book we will see in detail how human societies can be reorganised to align the interests of their government (management) with the interests of the members of the society. This will unleash the immense advantages of cooperation for the benefit of all.

6

The Evolution of Management

An organism that has the ability to control others could use its power to manage a group to establish complex cooperative organisation. It could, but would it? Does evolution favour organisms that use their power over others to promote cooperation between them? How could it be to their evolutionary advantage to use their power in this way?

Or does evolution favour organisms that use their power to take what they want off others? If evolution favours self-interest, will it not favour powerful organisms that plunder resources from others? Why manage, when you can just take? These questions are critically important for deciding whether the potential benefits of cooperation are capable of driving progressive evolution. If evolution cannot produce dominant organisms that use their power to promote cooperation, managed cooperative organisations are unlikely to arise, and evolution will be unable to progressively exploit the benefits of cooperation.

Part of the answer to these questions is found in an old saying: if you give a person a bag of corn you feed him once, but if you teach him how to plant and grow the corn, you feed him for life. In the same way, a powerful organism can either plunder a group of organisms once, or can manage and look after the group so that the powerful organism can harvest benefits from the group for life.

To develop these ideas more fully, we need to return to our generalised example of a population of organisms that are formed into groups. How will the population evolve when a dominant organism that is able to control others arises in the population?

Initially at least, the dominant individuals that would do best would be those that use their power to get the most food, the best opportunities to mate and produce offspring, the safest places to rest at night, and so on. Evolution will favour those that use their power purely for their own benefit, whether or not this has harmful effects on others.

Particularly if groups are able to accumulate food and other resources, evolution is likely to favour dominants that move from group to group, plundering resources as they go. In these circumstances, a dominant would not restrict the amount of resources it takes in order to limit the harm it does to a group. Any dominant that did so would be out-competed by more ruthless individuals.

But, in time, dominants may discover ways of getting other individuals to do useful things for them. They may discover how to use their power to get others to gather food, help defend the group against predators or other plundering dominants, or assist in producing and rearing the dominant's offspring. Dominants could do this by rewarding those that help them, or by forcing them to do so.

The discovery by dominants of ways to make others useful to them on an on-going basis completely changes the characteristics that evolution will favour in dominants. It changes the evolutionary forces that act on them. The dominants that will do better in evolutionary terms will now be those that stay with a particular group and harvest an on-going stream of benefits from it. Evolutionary success will go to dominants that are best at keeping their group healthy and productive so that it can continue to provide useful benefits. Plundering and moving on becomes a losing strategy.

The success of the dominants will increasingly depend on the success of those that they control. Evolution will favour dominants that are able to extract a stream of benefits from a group of individuals on an on-going basis. As dominants find more and more ways of making others useful to them, dominants will become more dependent on those they control. Increasingly the interests of dominants and the interests of their groups will be aligned. If a dominant harms those under its control, it harms itself. If it helps those under its control to live long and productive lives, it helps itself. The ability of the dominant to extract benefits from the group means that it is able to capture the benefit of anything it does to improve the capacity of the group to produce on-going benefits. A dominant will be as interested in the health and productivity of its group as a farmer is in his crops and animals. The dominants become managers.

It will be in the interests of a manager to keep its group healthy and productive. It will limit what it takes from the group so it does not damage the ability of the group to function and reproduce itself. And it will be in the interests of a manager to exclude other dominants from access to its group. It will do better by having the sole use of the resources and services that can be harvested from the group. So it will protect the group from being plundered by other powerful individuals or bands.

As managers become better at preventing other dominants from using their group, managers become more and more dependent on their group for their own evolutionary success. They will not be able to leave their group and move to another. If they damage their own group, they will not be able to escape the consequences of this by taking over another group.

Eventually, the evolutionary success of a manager will depend totally on its ability to manage the group that it controls. It will have no evolutionary future outside the group it manages. The success of the manager will depend totally on the success of the organisation. The manager will capture all the effects on the organisation of its actions, good and bad, and will therefore treat the organisation as it would treat itself. The self-interested manager will pursue the interests of the organisation as a whole. The group and the manager will form an organisation that is a single, united evolutionary entity.

Once managers develop the ability to extract an on-going stream of benefits from a group, they can also benefit from boosting the productive capacity of the group. And we have seen that the potential for doing this is immense. The productive capacity of a group can be boosted greatly if cooperation is organised within the group. We have seen that the barrier to the evolution of cooperation largely prevents un-managed groups from doing this. But the potential benefits of cooperation can be exploited within managed groups. A powerful manager can support cooperation and control cheating, theft and free riding. And a manager with the power to harvest benefits from a group will be able to profit from anything it does to improve the productivity of the group[1].

We can now see that not only is a manager able to unlock the immense benefits of cooperation, it is in its own evolutionary interests to do so. A manager that increases the productive capacity of its group by supporting cooperation will be able to harvest a greater stream of benefits from the group for its own use. A self-interested manager will therefore control its group to promote cooperation.

To summarise: when individuals that have the ability to control others emerge in a population, they will initially use their power in ways that often damage the interests of others. But once they discover ways to use their power to harvest a stream of benefits from others, it will increasingly be in their interests to protect and nurture a group of individuals that can provide these benefits.

Because dominants can harvest benefits from the group, they will be able to benefit from whatever they can do to increase the productivity of the group. The greatest potential for this lies in managing the group to overcome the barrier to the evolution of cooperation. It will be in the evolutionary interests of managers to make it in the interests of members of the group to cooperate and to serve the interests of the organisation as a whole. The result will be an organisation in which the interests of all, including the manager, are aligned with the interests of the organisation. The organisation will be cooperative even though it is made up of thoroughly self-interested components[2].

The evolution of early cells provides a concrete and specific illustration of this general evolutionary sequence. We begin with an autocatalytic set of proteins. As we have seen, an autocatalytic set can reproduce itself through time because each type of protein in the set helps make other members of the set. The proteins catalyse reactions of smaller organic and inorganic molecules progressively constructing larger and larger molecules, and eventually producing the members of the set.

A flourishing autocatalytic set of proteins would build up a rich soup of component molecules and other chemical processes that are used to produce the members of the set. But there is little to stop other large organic molecules from using this soup of component molecules and processes for their own reproduction. RNA molecules are particularly suited to this. Due to their greater size and stability, they are less able to be damaged by proteins, or to be stopped by proteins from using the components of the set. And they are better able to survive periods spent drifting between autocatalytic sets.

So self-reproducing RNA molecules could do very well by plundering the resources of an autocatalytic set to help their own reproduction, perhaps destroying the set in the process, and then moving on to another set to do the same again[3]. The RNA would have no interest in preserving the sets that it plundered.

As we have seen, this would change once RNA molecules discovered ways to manipulate the reactions within the set to their own advantage. For example, the RNA molecules might use their catalytic ability to boost reactions within the set that helped their own reproduction. The RNA would no longer just be plundering components from the set for its own use. Instead, it would boost reactions within the set that produced these components. So the RNA might boost reactions that produce proteins that help it to replicate more efficiently, or that help it to catalyse other reactions that are useful to it.

In this way, RNA molecules begin to discover ways to make autocatalytic sets useful to themselves on an on-going basis. The RNA begins to have an interest in ensuring that the set is not destroyed so that the set can continue to produce a stream of benefits for the RNA. The RNA will restrain its plundering of the set, and can benefit from a long association with it.

The RNA can further increase the stream of benefits it gets from its set by boosting the productivity of the set. It can do this by using its ability to control the set to overcome the barrier to the evolution of cooperation. As we have seen, the RNA can boost the production of cooperative proteins that make a significant contribution to the efficiency and effectiveness of the set, but that otherwise would be out-competed within the set. And it can suppress the production of proteins that free ride and cheat[4].

It will increasingly be to the advantage of the RNA to stay permanently with a particular autocatalytic set. This will enable it to harvest though time the benefits of the massive increase in productivity it has produced. And selection will favour RNA that prevents other RNA from plundering the set. The RNA can do this by managing the components of the set so that they cooperate together in ways that resist other RNA. For example, the RNA could manage the set so that the members cooperate together to produce a cell wall, or cooperate together to engulf and digest RNA that attempts to plunder the set.

This evolutionary sequence eventually produced proto cells in which particular RNA molecules lived permanently with a particular autocatalytic set of proteins encased in a cellular membrane or wall. Once proto cells emerged, neither the set nor the RNA molecules could have any life or future outside the proto cell. Their evolutionary success was totally dependent on the evolutionary success of the organisation as a whole.

But this complete alignment of interests between the RNA and the proto cell could break down when there was more than one RNA molecule in the proto cell. This is because the molecules would compete against each other, and the winner would not usually be the one that contributed most to the cell. An RNA molecule would do better if it invested its time and resources in promoting its own reproduction, so that its offspring increased in numbers in the cell and in the next generation of cells. It would out-compete cooperative RNA molecules that instead used their time and resources to boost the productivity of the cell, rather than to promote their own immediate reproduction[5].

This breakdown of cooperation was, of course, another instance of the barrier to the evolution of cooperation. And it could be overcome in the same way as other instances. The RNA molecules had to be controlled so that the only way that they could advance their own interests was by advancing the interests of the organisation as a whole. They had to be prevented from gaining any competitive advantage over other RNA, except where this was because they had contributed more to the cell[6]. Once this was achieved, cells could include many RNA molecules, with each molecule specialised to manage some particular aspect of the chemical and other processes of the cell.

How was destructive competition between RNA prevented within cells? As for the suppression of competition amongst mitochondria, an important development was to prevent RNA moving from one cell to another. This was accomplished in large part by the evolution of a cellular membrane. The membrane physically prevented RNA from plundering the cell and then leaving, and tied their interests to the interests of the cell[7]. A further step in suppressing destructive competition was the joining together of the RNA molecules to form a single chromosome[8]. All the genetic material was bound together, and could not reproduce separately. Any particular RNA molecule was unable to produce more offspring than others. What one molecule did, the others did as well. Ensuring all RNA molecules shared the same reproductive fate prevented competition. They were all in the same boat.

But this did not solve the problem completely. If chromosomes could reproduce freely within the cell, producing a number of chromosomes, competition could resume. The chromosomes could compete with each other within the cell. This was prevented by controls over the reproduction of chromosomes. The controls ensured that a chromosome could reproduce only when the cell reproduced, and that only one copy of the chromosome would go to each daughter cell. So when the cell split into two daughter cells, each chromosome would also produce two daughter chromosomes, one going to each daughter cell. Chromosomes were prevented from reproducing at any other time within the cell. As long as these controls were successful, competition could not arise.

Modern cells have evolved elaborate structures and processes that strictly control the reproduction of the genetic material. Genes are joined up into chromosomes, and in the processes of mitosis and meiosis, the reproduction of chromosomes is synchronised with the reproduction of the cell. Chromosomes are strictly controlled so that none can gain a competitive advantage by getting into more daughter cells than others, or by making more copies than other chromosomes[9].

The result is that there is only one way a particular RNA or DNA gene can get a competitive advantage over any other. It must help the cell that contains it to do better than other cells that contain alternative genes. The interests of a gene are completely aligned with those of the cell that contain it.

Once RNA molecules were controlled in these ways within early cells, destructive competition was suppressed. But how were these controls actually established? What did the controlling? What had the power to control RNA molecules?

The answer is that individual RNA molecules could be controlled by the actions of the cell as a whole. That is, they could be controlled by structures and processes produced by the autocatalytic set managed by RNA. The organisation as a whole had the power to control any individual within it. So RNA molecules, in conjunction with the parts of the cell they managed and organised, were able to produce structures and processes that controlled individual RNA molecules, including themselves. And if individual molecules were ever able to find a way to manage part of the cell to break out of these controls, it would be in the interests of other RNA molecules to respond by managing the cell to reinstate the controls[10].

A broadly similar evolutionary sequence also contributed to the emergence of the cooperative human societies managed by powerful rulers that began to develop about 10,000 years ago. With the discovery of agriculture, and the establishment of settled tribal communities that accumulated food and other possessions, bands of individuals could make a living by moving from community to community, plundering food and possessions as they went. Initially, these powerful bands would have simply taken what they wanted, without any interest in the effects this might have on the capacity of the communities to continue in existence.

But the success of this strategy would decline as the number of bands and the rate at which they plundered communities increased. A band might then find that it could do better by taking over a community permanently, protecting it from other bands, and using its power to extract a stream of benefits from the community on an on-going basis. These benefits would not only include food, weapons, labour and sex. A band could also reap psychological benefits such as the experience of exercising power over a community, and feelings of self-importance and superiority.

The band would now have a direct interest in doing anything that could boost the productivity of the community. This would enable it to increase the stream of benefits it harvested. As we have seen, the greatest potential for improvements lay in the band using its ability to control the community to promote cooperation. The band could do this by aligning the interests of the individual members of the community with the community interest, and therefore with its own interests.

For example, the band could tax the community to fund activities that were beneficial to the community as a whole, such as irrigation and defence. And the band could establish an effective framework of controls to support reciprocal exchanges between community members. It could do this by punishing cheats and free riders, enforcing exchange agreements, and settling disputes fairly. In this way, a community that farmed crops and animals would in turn be farmed by a band, and the band would become a manager[11].

The better the band became at managing the community, the more powerful it would become. It could increase its power further by taking over other communities and by managing them on an on-going basis. As well as enabling the band and the community to defend better against others and to harvest greater material benefits, this would also increase the psychological benefits that could be harvested by the band and its leader. They could surround themselves with the trappings of power, and could even ensure they were treated as gods.

After a number of generations, the person who led the band and who ruled the community was a king, or even an emperor.

Just as proto cells had to find ways to stop destructive competition between RNA molecules, kingdoms had to find ways to stop competition amongst powerful individuals within the society. The simplest way was to have a supreme ruler who had power over all others in the society. It would be in the interests of the ruler to control all less powerful individuals to ensure that they acted in the interests of the society.

But the possibility of competition for the position of ruler would arise again whenever a ruler died. Other individuals could then compete to take over the position. Some might even hasten the death of the ruler to open up this possibility. The scope for this competition would be reduced if the number of people who were qualified to take the position of ruler were limited. For example, if only children of the existing ruler were permitted to become the ruler. This greatly reduced the numbers of individuals who had the supreme incentive to overthrow the ruler.

However, this method of controlling destructive competition has had limited success. History is full of examples of sons killing fathers, wives killing husbands, and brothers killing brothers in order to meet whatever restrictive criteria were set to determine who could take the position of ruler[12]. The result has been as destructive to the proper management of these human societies as competition between RNA molecules was to the interests of proto cells.

The solution was the same as in cells: the ruler had to manage the organisation to establish processes and structures that could control even rulers[13]. Laws and methods of enforcing laws had to be established that could prevent rulers and those who could become rulers from seriously damaging the society. The effect of these laws and enforcement processes would be to move toward realigning the interests of rulers and potential rulers with the interests of the society as a whole. As we shall see, the establishment of controls over rulers (including governments) so that their interests were aligned more closely with the community has been very important in the evolution of human society, and will continue to be so[14].

Provided there was strong competition between kingdoms, the interests of the ruler and the community would be closely aligned. In these circumstances, the king would have to ensure that the community was well governed if the kingdom was to survive external threats and if he was to remain king. Only the best-managed kingdoms would survive. The better the ruler was at promoting cooperative relationships between citizens and at establishing cooperative activities that benefited the community, the more secure he and the kingdom would be.

But without strong competition, the interests of the governor and the governed would not be so closely aligned. If the kingdom had no external threats, there would be no immediate cost if, for example, the ruler and his allies over-taxed the community. The survival of the kingdom would not be threatened immediately if the ruler overindulged himself and his allies with material and psychological benefits at the expense of the community. Although the ruler would still be heavily dependent on the community and their interests would still overlap somewhat, the ruler could get away with over-exploiting the community.

In the evolution of proto cells, competition between cells ensured that the interests of the RNA and the interests of the autocatalytic set remained very closely aligned. This competition produced natural selection between cells and ensured that only the best-managed cells could survive and be successful. But in human evolution, strong competition between human societies has become less significant as the societies have increased in scale. Unlike cells, there is no longer a large population of small human social groups that are continually competing against each other for survival. Natural selection no longer operates strongly between human societies.

But as competitive pressures have declined, rulers and their allies have not had a completely free hand to exploit human societies. It is not in the interests of the average member of these societies to be exploited. It is in their interests to stop a ruler who attempts to do so. Although the rulers and their allies have much greater power than any individual, if enough individuals band together, they have sometimes been able to develop sufficient power to overthrow such a ruler. Collective action of this sort, and the threat of it, have been very significant in maintaining some alignment of interests between the governor and the governed in human societies as competition declined.

But when collective action successfully overthrows a ruler, it strikes a major problem. The society must continue to be managed effectively if it is to be successful. However, the collective itself is unable to govern the society effectively. It has little option but to put its own leaders into power, to make them the rulers.

The challenge then becomes: how do you establish a system of controls that makes sure that the interests of the new managers are permanently and irrevocably aligned with the interests of the society? How do you ensure that it will not be in the interests of the new managers to behave in the way the previous managers behaved? What controls will ensure that a self-interested manager will act only in the interests of the society as a whole? This challenge is of critical importance. If the interests of the new rulers lie in the same direction as the previous rulers, the same behaviour can generally be expected. Only new controls can permanently change the way a society is governed.

History is overflowing with examples in which revolts and revolutions have failed to devise controls that meet this challenge. Instead they produced new rulers whose self-interest lay in exactly the same direction as the overthrown rulers. The pigs took over the farm, and eventually become indistinguishable from the farmers they replaced[15].

However, there have also been examples in which new controls have been put in place that help to better align the interests of the governors and the governed. The democratic process is a clear example. In theory at least, democracy allows the governors to be overthrown by popular vote if they do not serve the interests of the society.

How do our present systems of government rate against the evolutionary ideal in which the interests of management are fully aligned with those of the managed? Are the interests of those who exercise power in democratic government completely aligned with the interests of the social systems they manage? Do the structures and processes that control their interests guarantee this?

This issue will be critical to the future evolutionary success of human organisation. To be competitive in future evolution, human organisation must comprehensively get around the barrier to the evolution of cooperation. If human society is to be best placed to participate in future evolution, it must be able to fully exploit the immense potential benefits of cooperation in the interests of the society as a whole. We can be sure that in the future, as in the past, organisations that are best able to exploit the benefits of cooperation will be the most competitive. They will be best at meeting whatever evolutionary challenges arise. They will have the greatest capacity to adapt successfully to the widest range of possible future events.

For continued evolutionary success, human society must be able to establish cooperation wherever and whenever it is beneficial to society as a whole. As we have seen, this can be achieved if a system of government aligns the interests of the individual members of society with those of the society as a whole. This will ensure that individuals who pursue their own interests will pursue the interests of the society. But government will do this only if the interests of the government are completely aligned with those of the society. When this condition is met, governments that pursue their own interests will also pursue the interests of the society. They will use their power to organise cooperation that serves the interests of the society.

Where the interests of government are not aligned in this way, the society will not be able to fully exploit the benefits of cooperation to satisfy the needs of its members. This would be the case for a society in which the destiny of governments is controlled by the citizens who have the greatest economic power. In such a society, it would be in the interests of the wealthy to have the government manage the society in their interests rather than in the interests of the society as a whole. Many believe that our current social systems are in fact organised in this way: in a 1992 survey of US citizens, 80 per cent said that they thought their country was run for the rich[16]. But our present democratic governments and the economically powerful tend to disagree with this. They generally argue that there is little need to radically change our social systems. They are vigorous supporters of the status quo. They claim that our current democratic systems are, in the broad, the best available forms of government. Democratic government is as good as it gets.

But throughout history, rulers, governments and those who benefit most from them have always extolled the virtues of the particular form of government that provided them with their wealth, power and self-esteem. They often claim that their system has been designed in accordance with God's will, or even that they themselves are gods, or at least that they have God on their side. And they commonly promote ideologies, myths and world views that support their system as the ultimate form of government, as the culmination of human history, and as the best and fairest method of organising human society. Any who push for alternative forms of government and who are sufficiently successful to threaten the established order have commonly been killed, jailed, beaten, or treated as criminals. They are often branded as mentally unstable, idiots, sexual deviants, ignorant and unscientific, or as being in the pay of enemy governments[17].

There are no examples from history of governments that have taken an evolutionary approach to systems of governance. Governments have never seriously promoted the view that they and their systems are temporary, merely a transitory step along the way to new and better forms of governance. Universally they have failed to demand that society should continually be searching for improvements in government, and should establish systems of government that are designed to evolve and to be better at evolving.

All this is exactly what would be expected of managers who have been constructed by evolution to pursue their own interests, and who have not been subject to controls that align their interests with those of the organisation they manage.

As we will see in detail, our current systems of government fail to meet the evolutionary ideal. Democratic processes are not sufficient to completely align the interests of the governors with the governed. As a result, even highly democratic human societies do not harness the benefits of cooperation in the interests of society as a whole[18]. A major challenge for humans in the future is therefore to devise forms of organisation that overcome this limitation, and that are more likely to be successful in evolutionary terms. Even though our societies may not be subject to strong and continual competition at present, we must organise our societies cooperatively so that they will be competitive in future evolution. If evolution on this planet is to continue to progress by exploiting more of the potential benefits of cooperation, we must meet this challenge successfully. In Chapter 17 we will look in detail at how humanity must reorganise its economic and social systems if it is to participate successfully in future evolution.

The origin of life itself provides a final illustration of how a system of management can emerge and exploit the benefits of cooperation. We begin with a watery environment early in the history of the earth. As the famous experiments of Miller and others[19] have shown, under the conditions that existed on the early earth, small organic molecules were produced in the atmosphere and elsewhere, and then accumulated in pools and other bodies of water. Reactions resulting from chance interactions between these smaller molecules produced a variety of larger organic molecules in the pools.

It is of critical significance for the evolution of life that some of these larger molecules had the ability to catalyse reactions amongst some of the smaller molecules. They had this catalytic ability because they were able to control the movements and behaviour of some of the smaller molecules. Large organic molecules are able to exercise control over smaller ones in the sense that they can collect them from a solution and hold them in particular positions, causing the smaller molecules to react. Importantly, the large molecules are not destroyed or modified when they do this. They continue in existence, able to repeat the process, organising more reactions of smaller molecules. In large part they can do this simply because of their greater size and stability. They can push other molecules around without reacting with them.

In this way, some larger molecules would coordinate reactions that were very unlikely to occur by chance in the mixture. These reactions could proceed only when particular combinations of smaller molecules came together at the same time in a particular spatial pattern. The likelihood of this occurring by chance would be very small in the absence of the management applied by the larger molecules.

So large organic molecules were likely to emerge that could manage smaller molecules to organise reactions that would not otherwise occur. These reactions might produce another larger organic molecule, or they might break down a molecule, releasing energy to power other reactions. New types of larger molecules would be formed in the mixture, and these in turn would manage smaller molecules to produce further new types of larger molecules. In this way, management by larger molecules opened up new possibilities for the mixture and for evolution. It generated new types of larger molecules with new properties that would not have emerged in the mixture otherwise.

The content of the mixture would change through time as a result of these processes and as conditions changed. But these changes would not unfold in any particular direction. Whatever smaller molecules a large molecule managed, and whatever reactions it produced would make no difference to whether the large molecule survived. The larger molecules would not compete with each other on the basis of what they managed. A particular change in what a large molecule managed would not make any difference to how many of the large molecules would be formed in the mixture, or how long they survived. Natural selection would not operate. It would apply only if some forms of management did better in the mixture than others. Changes in what was managed by larger molecules would be more or less random.

But all this changed suddenly if the mixture generated a large molecule that happened to have a critically important property. This property was the ability to manage other molecules to produce a reaction that boosted the rate at which the large molecule itself was formed. Suddenly, there would be a form of management that reinforced its own existence. Management would have hit upon a way to use its power to help make other instances of itself[20].

This would also mean that a form of management had been discovered that could do better than alternative forms of management. If the change in the larger molecule produced management that was more effective at boosting the reaction that produced itself, it could out-compete other varieties of large molecule. It would be better at using the resources of the mixture to reproduce itself. Natural selection would now apply between different varieties of large molecules, and this natural selection would select large molecules that were better at managing the production of other instances of themselves.

The large molecule could boost its own reproduction either directly or indirectly. The boost would be direct if the reaction managed by the large molecule produced another instance of the large molecule itself. For example, the large molecule might act as a template for a copy of itself, as in the self-replication of RNA and DNA molecules. The boost would be indirect if the managed reaction produced other molecules that in turn directly or indirectly boosted the production of instances of the original large molecule. If a collection of different types of large molecules each indirectly boosted their own reproduction in this way, the result would be an autocatalytic set that reproduced as a whole. The set as a whole would manage matter to produce new instances of all its key components. As we have seen, there is good reason to believe that the emergence of an autocatalytic set in a sufficiently rich mixture of organic molecules might be easy and not at all improbable[21].

When a self-reproducing manager first arose in a rich mixture of organic molecules, it would tend to take advantage of the organic molecules that had built up in the mixture. It would plunder the rich capital of molecules that had accumulated in the mixture, using them as food and as components for its reproduction.

But the same sort of evolutionary sequence that turned RNA and human plunderers into farmers could also be expected to have applied to the evolution of autocatalytic sets. Sets would do better when they discovered ways to manage and boost the chemical process that produced the molecules that they previously plundered. Instead of just taking the molecules they needed from the mixture, they would begin to manage the processes that produced the molecules. They would use their management capacity to enhance and boost these processes, harvesting the products for use in their own reproduction. And they would discover ways to manage these chemical processes so that the processes cooperated together through specialisation and division of labour. The result would be a proto metabolism of interlinked chemical processes that were permanently managed by the set to produce energy and material for its reproduction[22].

In this Chapter, we have seen that evolution tends to produce managers whose interests are aligned with the interests of the organisations they manage. It is in the evolutionary interests of managers to use their power to promote and support cooperation within their organisation. This enables them to harvest greater benefits from their organisation, and ensures that they and their organisation are more competitive. In this way, evolution can overcome the barrier to the evolution of cooperation. It can build cooperative organisations out of self-interested components. Once evolution has exploited some of the potential benefits of cooperation by building cooperative organisations, it can progress further by building cooperative organisations of those organisations. And then build cooperative organisations of organisations of organisations. In this way, evolution will progressively exploit the benefits of cooperation by building cooperative organisations of wider and wider scale.

But to complete our investigation of how evolution can progress by organising cooperation, we must consider another form of cooperative organisation. In this Chapter and the previous Chapter we have seen how an external manager can organise cooperation in the group it manages. In the next Chapter we will see that there is another form of management, internal management, that can also organise a group cooperatively.

7

Internal Management

Complex cooperative organisations can be organised by a powerful manager that is external to the individuals it manages. Management is obviously external in all the cases we have considered to this point: atoms and small molecules were managed by larger molecules to start the evolution of life; autocatalytic sets were managed by RNA to form early cells; mitochondria were managed within these cells to produce modern eukaryote cells; and human social systems were managed by external rulers and governments to form modern human societies.

But not all complex cooperative organisation includes a separate manager that supports cooperation and controls cheats and free riders. For example, multicellular organisms do not include a separate manager that controls other cells within the organism[1]. Beehives, ant colonies and other insect societies also do not appear to include a dominant individual that uses its power over others to organise cooperation[2]. And before the rise of managed agricultural communities about 10,000 years ago, small bands of humans cooperated in their hunting and other activities without any control by a separate ruler[3].

How has the barrier to the evolution of cooperation been overcome in these cases? How are cheating, theft and free riding prevented, and how are resources redistributed within the organisation to support cooperation? Has this been achieved without the use of power and control, or is the source of control just less obvious in these cases, compared with where the manager is separate?

We will see that these organisations without separate managers are able to evolve complex cooperation only because they are managed. But the source of the management is not external to the individuals being controlled. It is internal to each of them[4].

To see how this internal management can operate, consider a tribe of humans that includes a number of individuals who specialise in tasks useful to the tribe. Someone makes arrows and spears, another makes bows, others specialise in hunting, and so on. These specialists survive in the tribe by exchanging what they produce for the necessities that they do not produce. But as we have seen, cheating can undermine this cooperation. If cheating is not prevented, the specialists will not be able to survive as specialists, even though their cooperative division of labour is extremely beneficial to the tribe.

Cheating could be prevented in the tribe in either of two ways: first, cheating would not be profitable if a powerful tribal ruler punished individuals who undermined cooperation by cheating. Second, cooperation would not be disrupted by cheats if the tribe was made up of individuals who each had a genetic predisposition to be trustworthy and honest. Cheating would not occur if each individual carried genes that prevented him from being a cheat. For example, the genes might make the individual feel emotionally repelled by the idea of acting dishonestly. It would not be in the nature of such an individual to cheat.

The same result would be achieved if, instead of a genetic predisposition, all members had a learnt predisposition against cheating. There would be no cheating if all members of the tribe had been inculcated with a very strong belief that they should be trustworthy and not cheat in exchanges. For example, they might have been taught that cheating was fundamentally wrong, not the way for any respectable member of the tribe to behave, and against the wishes of the gods who created the tribe and its world. Norms, moral beliefs and other rules of behaviour that are supported by religious systems are examples of learnt predispositions that would be capable of controlling behaviour in these ways.

To control the members of the tribe successfully, the genetic or moral predispositions would have to be sufficiently strong to stop the individuals from doing what would otherwise be in their self-interest. Even when an individual could see that cheating was clearly in its interests, its predispositions would have to stop it from doing so. They would have to be strong enough to cause individuals to resist temptation successfully. Individuals would have to be left with no choice. Ordinary beliefs that changed as circumstances changed would not be sufficient. Beliefs that could not be changed by the individual such as sacred beliefs would be necessary. Strong genetic or learnt moral predispositions of this kind could control the tribe as completely as a powerful external manager. The tribe and its members would be hard wired to behave cooperatively.

To control all the members of the tribe, the hard-wired predispositions would also have to be contained in each and every member of the tribe. Any individual that escaped the hard wiring would not be restrained from pursuing its individual interests at the expense of the group. It would gain the benefits of being in a tribe of cooperators without having to contribute to the cooperation. It would be free to cheat and free ride, and would advance its own interests by doing so. If many members of the tribe did not contain the hard wiring, cooperation would be undermined, and cheating and free riding would take over.

A further condition would also have to be met if the control was to be maintained in a tribe over time: the hard wiring would have to be produced in all the members of the tribe across the generations; through time, the tribe would have to contain only individuals who included the hard wiring. This would be achieved if the hard wiring was produced in each new individual who was born into the tribe. The normal process of genetic inheritance would tend to do this for genetic predispositions. Hard wiring that was acquired by learning could also meet this condition if it included a predisposition to inculcate others with its beliefs and predispositions. For example, the learnt hard wiring might cause parents to inculcate their children with a set of moral and religious beliefs. One of these beliefs could be that parents must teach the entire set of beliefs to their children. If the inculcation was successful, the children would grow up to teach their children the set of beliefs, including the belief they should repeat the inculcation process. Each generation would inculcate the next with the set of beliefs. Each would teach the next the difference between 'right and wrong'.

A religious cult within modern society is an example of a set of learnt predispositions that usually include a predisposition to inculcate the beliefs in others. The set of predispositions organise the members of the cult to behave in ways that help the predispositions to reproduce themselves in others. The cults that flourish are those that are based on a set of beliefs that are best at organising their own reproduction.

The processes of inheritance and inculcation would go a long way to ensuring that genetic and learnt predispositions were reproduced in all members of the tribe. But they would not be 100 per cent successful. Mutations would arise from time to time in both genes and in moral beliefs. However, such breakdowns in hard wiring could be repaired. The predispositions could be maintained universally in the tribe if those who were not hard wired to cooperate were punished and perhaps expelled from the tribe.

This punishment could be organised by the hard wiring itself. If this were the case, the hard wiring would not just include predispositions to cooperate in various ways and to refrain from cheating. The hard wiring would also include a predisposition to punish and expel members of the tribe who acted as if they did not include the full set of hard-wired predispositions (including the predisposition to punish). So individuals with the hard wiring would cooperate, and they would punish individuals who did not cooperate, or who failed to punish cooperators[5].

For example, where the hard wiring was genetic, it might cause members of the tribe to be emotionally outraged by the behaviour of non-cooperators and to be impulsively aggressive toward them. Where the hard wiring was learnt, members of the tribe might be inculcated with a strong moral belief that non-cooperators were deviants who deserved the strongest punishment because they had set themselves against all that was valued by the tribe and its gods.

In this way, a tribe could be managed by a set of hard-wired predispositions that could reproduce itself in most of the members of the group, and cause them to expel any remaining members who did not appear to contain the hard wiring. I will refer to such a manager as a distributed internal manager: the management is internal to the individuals being managed, and is distributed in the sense that it is contained in each of the members of thc group.

A distributed internal manager is formed of a set of hard-wired predispositions that are reproduced in each of the members of a group of organisms. As we have seen, the predispositions can be genetic, or can be strong moral or other beliefs that are learnt during the life of the organism. As we shall see, internal managers are established by genes in insect societies and in the societies of cells that are multicellular organisms. And they were established by systems of moral and religious beliefs as well as genetic predispositions in the early tribal societies that preceded modern human societies managed by external rulers.

A distributed internal manager clearly has the potential to organise a group cooperatively. It can hard wire individuals so that they participate directly in cooperation, provide resources to maintain group activities, refrain from cheating, theft and free riding, and punish individuals who undermine cooperation. In this way it can cause individuals to act in the interests of the group as a whole, and to treat others in the group as self[6]. An internal manager can organise cooperation directly by hard wiring specific cooperative behaviours in individuals. Or it can do it indirectly by controlling exchanges between individuals so that individuals capture the benefits of their cooperative acts on others. For example, individuals might be hard wired so that they do not cheat in reciprocal cooperative exchanges. This would enable the reciprocal altruism mechanism to operate successfully.

Internal management has the potential to organise cooperation in these ways, but will evolution produce management that does this? Will evolution favour the emergence of distributed internal management that organises a group to act cooperatively? More specifically, will cooperators organised by internal management out-compete non-cooperators within a population or groups of organisms? It turns out that just as evolution favours external managers that support cooperation, it will also favour internal managers that do so. This is because the manager is able to capture any benefits that it produces, and cooperation is capable of producing immense benefits.

There are a number of ways in which an internal manager can use its power to capture the benefits of any cooperation that it organises. As we have seen already, the manager can organise the individuals it controls to exclude non-cooperators, cheats and free riders from their group. If all individuals who do not contain the manager are expelled from a group, the manager will capture all the benefits it organises within the group. This is because each and every member of the group will contain a copy of the manager. So if an individual helps others in the group, it helps copies of the manager. The manager will also capture the harmful effects of any behaviour that it causes within the group. If an individual hurts others in the group, it hurts the copies of the manager that they contain. If the sum total of the effects of an action on others are harmful, the total effects on the manager will be harmful. And if they are beneficial, the manager will benefit overall. As a result, if the manager can cause an individual to act in ways that produce net benefits for members of the group, the manager will benefit, even though the behaviour might not benefit the individual. The manager will capture the benefits of whatever it can do to advance the position of the group. The evolutionary interests of the manager will be aligned with the evolutionary interests of the group as a whole. What is good for the group will be good for the manager.

There are other ways in which a manger can ensure that it captures the benefits of any cooperation that it creates. It can organise the individuals that it controls to direct their cooperation only toward others who contain copies of the manager. This will ensure that only individuals who contain a copy of the manager will benefit from the cooperation. For example, if the manager is a set of norms and moral codes, it will prevent individuals from cooperating with others who do not follow its codes and norms. If the manger is genetic, it will prevent individuals from cooperating with others who are unlikely to contain copies of itself, such as non-relatives. Cooperators organised by such a manager will be able to out-compete non-cooperators within a group or within a population. The manager may begin by controlling only a few individuals, but eventually it will take over the entire group or population.

As with external managers, the best way in which an internal manager can profit from its ability to control a group is by promoting cooperation. And the better a manager is at organising cooperation, the greater the benefits it can capture. If a number of managers are competing within a group, the manager that is best at organising cooperation will do best, provided it can capture the benefits of the cooperation it organises. If a number of groups are competing, the group with the manager that is best at organising cooperation will do best. Both within groups and between groups, evolution will favour managers who use their power to overcome the barrier to the evolution of cooperation. Evolution will produce distributed internal managers that use their power to establish complex cooperative organisations. We have seen already that evolution can progress by producing cooperative organisations managed by external managers. It can also progress by using internal managers to organise cooperation. As we shall see in detail in Part 4 of this book, evolution on this planet has used both internal and external management to progressively produce cooperative organisations of larger and larger scale.

The most successful internal managers will be those that are best at capturing the benefits created by their management. We have seen that some of these benefits will be lost to the manager if it is unable to suppress all cheats and free riders in the organisation. In general, benefits will be lost if any members of the organisation contain a different manager. And it is not just that these other managers will capture some of the benefits of cooperation. It will also be in their interests to compete within the organisation to extract whatever benefits they can get. Competition between different managers within an organisation will seriously undermine cooperation. As a result, evolution will favour managers that are better at suppressing competition from other mangers and at preventing them from taking any of the benefits of cooperation. It is no accident that the systems of moral beliefs that have survived to the present are usually highly intolerant of those with different beliefs[7].

Distributed internal management that is unable to suppress competition from other managers within the organisation will not be effective at organising cooperation. It will not be able to capture all the benefits of cooperation. An example is the process of genetical kin selection that we considered earlier. Kin selection begins to operate when an individual contains a genetic manager that predisposes it to cooperate with others. The normal process of reproduction will produce related individuals who also contain the same manager. If it is to be successful, the manager must capture the benefits of the cooperation it produces. So evolution will favour genetic managers that organise individuals to direct their cooperation only towards others who contain copies of itself. As we have seen, managers can use relatedness to target their cooperation in this way[8]. But, because relatedness is not a perfect indicator of a shared manager, a manager will not capture all the benefits of the cooperation it organises. Some of the benefits will leak to individuals who do not include the manager. The manager will not find it profitable to organise some forms of cooperation even though it provides net benefits. As a result, the ability of genetical kin selection to organise cooperation is limited.

Genetical managers in some organisms have found better ways to capture the benefits of their management, and are better at establishing cooperation. The genetical managers of white-winged choughs provide an example. White-winged choughs are amongst the most cooperative of birds. All choughs in a group can be involved in feeding the young produced by the group, whether or not they are closely related. But although the members of the group might not be close relations, they are all likely to contain the genetical manager that produces this cooperation. And the manager will therefore capture the benefits the cooperation produces. This is because the manager organises the members of the group to punish and expel those who do not cooperate. Individuals who contain a different manager that causes them to attempt to free ride or cheat will be forced out of the group[9].

The internal genetic managers of the most complex insect societies have developed mechanisms that are even more effective at ensuring they capture the benefits of their management and suppress internal competition from other managers. The managers organise their society to increase the probability that only individuals that include the manager will contribute to the reproduction of the colony[10]. To the extent they are successful, internal competition will be prevented, and cooperation will not be undermined.

Managers attempt to do this by organising the society so that it can reproduce only through the sons and daughters of the queen who originally founded the colony. As far as possible, all other members of the society will be prevented from producing offspring that can found a new colony. Managers have discovered numerous ways of organising a society to help achieve this. For example, in most ant colonies, queens are predisposed to produce a pheromone that suppresses egg production in worker ants[11]. The workers themselves are hard wired to respond to the pheromone in this way. Queens will also attack any worker ants that develop ovaries, preventing them from reproducing[12]. And in honeybees, workers are predisposed to eat any eggs that are laid by other workers[13].

If the manager is able to successfully organise the society in this way, any individual who does not contain the manager will have no opportunity to compete with the manager to contribute to the next generation. Destructive competition within the society will be suppressed. Within complex insect societies that include these management controls, there is little competition and conflict. But internal conflict is common in smaller, less complex insect societies, particularly where more than one manager can reproduce and found a new colony[14]. Within these societies, the competing managers will attempt to organise the individuals they control to preferentially direct their cooperation toward the other members of the society who are most likely to include a copy of the manager.

In these examples, evolution has produced arrangements that suppress competition from other individuals and from other managers within the organisation, enabling cooperation to be established. But this suppression of competition amongst alternative managers produces another evolutionary problem. Evolution proceeds by producing and testing new arrangements that may prove to be better than the old. If no new possibilities are tested, evolution stops, and organisations are unable to discover new and better adaptations. Management can evolve and improve only by testing out new forms of management. But if individuals with mutant management arise within an organisation, they can compete with the existing manager, taking resources and undermining cooperation. This is similar to the destructive cooperation that can arise between RNA molecules within proto cells. It will be in the evolutionary interests of mangers to quickly suppress any new managers that arise within the organisation. But this will cause another problem—the suppression of new managers will also suppress evolutionary innovation.

In the case of the tribes of humans that were managed by sets of strongly held beliefs and other behavioural predispositions, this meant that new ideas and innovative beliefs could not be tolerated. Any deviant behaviour that conflicted with the set of moral beliefs that managed the group would be vigorously suppressed. Groups that were organised in this way would therefore have very stable systems of belief, but at the price of a stunted ability to adapt and innovate. The beliefs that managed the group would be very conservative, and disagreement would be strongly punished. As we will see later when we deal with human evolution in detail, this is a central reason why human societies managed by separate rulers have been far more successful in evolutionary terms.

But where distributed internal management has been able to establish complex cooperative organisation, it has done so because it has discovered ways to suppress internal competition without also suppressing innovation. The best example is multicellular organisms, which are organisations of cells managed by distributed internal management. Each cell generally contains a copy of the same genetic material. These genes can control the actions of each cell within the organisation, organising whatever cooperation is beneficial. This internal management has established the extraordinary level of cooperative specialisation and division of labour found within multicellular organisms, including ourselves.

Managers in complex multicellular organisms have discovered two sets of arrangements that combine together to suppress competition without putting a stop to evolutionary innovation. First, managers organise the development of the organism so that only a small group of cells, the germ cells, can produce the eggs or sperm that reproduce the organism. All other cells and their managers are locked out of reproduction. This is achieved early in development when the germ cells are separated out from the other cells of the organism, and are kept physically apart. So the overwhelming majority of the cells of the organism (and their managers) are prevented from competing to get into the next generation. There is no practical way for them to produce eggs or sperm to participate in reproduction. This greatly reduces the possibility of destructive competition between germ cells and other cells within the organism[15].

The potential for competition between germ cells is also suppressed. In principle, they could compete with each other to form the most eggs and sperm. But the germ cells are kept inactive and prevented from reproducing. They are therefore largely powerless within the organism—any mutant genetic manager that arises within the germ cells is unable to produce a group of cells that it could manage to pursue its competitive interests against other germ cells.

The second way in which competition is suppressed is to have the multicellular organism reproduce by passing through a single cell stage formed by the union of egg and sperm[16]. This single cell then divides repeatedly to form a new multi-celled organism. So only a single manager gets into a new organism. This greatly reduces the potential for disruptive competition: the only way new managers can arise in the new organism is through mutations that occur after the bottleneck of the single cell stage. And, as we have seen, competition amongst any fresh mutants is suppressed by separating out germ cells from other cells early in development.

These two sets of arrangements combine to enable new mutant managers to be produced and tested without destructive competition within organisms. Mutant cells and mangers are produced in the germ cells, but cannot compete destructively there. Neither can they compete with each other when a new organism is formed. The single cell bottleneck ensures that only one manager gets into each new organism that is produced. Fresh mutants could arise as the new organism grows, but competition from these is suppressed because they are prevented from reproducing. As a result, destructive competition within the organism is largely prevented. New mutant managers can arise, but the only way they can compete with other managers is through competition between the organisms that they manage. The manager that organises the best organism will win. So multicellular organisms are able to try out new and possibly better managers without unleashing destructive competition within the organism.

The evolution of insect societies and of multicellular organisms illustrate how internal management can control and organise a group of living processes to suppress competition and promote cooperation. It can be as effective as external management. But internal distributed control is not as visible or as obvious as external control. This is not just because it controls individuals from within. It is also because the control is distributed amongst each of the individuals that are being controlled. The operation and evolution of internal management cannot be understood by looking at only the behaviour and the evolutionary fate of just one individual. The behaviour and the evolutionary fate of all of the individuals who contain the manager must be taken into account. An internal manager organises its group as a coordinated whole, and it can be understood only as a whole. For example, it is not possible to understand the evolutionary significance of moral codes or systems of religious beliefs without recognising their group effects.

A group that is managed by distributed internal management is as much a vertical organisation as a group managed by a separate, external manager. It has an additional layer of organisation compared with a group that is not controlled by a manager. This additional layer is the set of predispositions that are reproduced in each of the members of the group through time. These predispositions control and coordinate the behaviour of members of the group. In the examples we have considered, the additional layer can be a set of genetic predispositions or a set of learnt beliefs such as moral codes.

The group cannot be said to include such an additional level of organisation unless a common set of predispositions is reproduced across the members of the group. If each of the individuals in the group has a different set of predispositions, there is no manager that controls and coordinates their behaviour as a group. Each individual will be controlled by its particular set of predispositions, and these will be a distinct level of organisation within the individual. But the group as a whole is not controlled and coordinated unless a common set of predispositions is reproduced across the group. Different sets of genetic or moral predispositions in different individuals do not create a new level of organisation across a group.

*   *   *   *   *

This brings us to a point where it is useful to summarise where we have got to in completing the task I set myself in Chapter 2. In that Chapter we identified what I would have to do to show that the evolution of life progresses toward increasing cooperation. How far have we got?

In Chapter 3 we saw that cooperation between living processes can be extremely advantageous in purely evolutionary terms. The benefits include specialisation, division of labour, the discovery and exploitation of new types of adaptation, coordinated adaptation over wider scales, and suppression of destructive competition. Potentially, these benefits will be produced by cooperation between any living processes, wherever they are in the universe. As a result, the benefits of cooperation will not be fully exhausted until all living processes participate in the one cooperative organisation. Until this occurs, the unexhausted benefits will continue to drive evolution that progressively exploits the benefits.

However, we saw in Chapter 4 that evolutionary mechanisms including gene-based natural selection cannot easily exploit the benefits of cooperation. There is a barrier to the evolution of cooperation that applies generally to all living processes, including to self-reproducing molecular processes, cells, multicellular organisms, and societies of organisms. Evolution favours individuals who pursue their own evolutionary interests. In general, individuals will not cooperate unless it pays them to do so. So despite the benefits of cooperation, evolutionary mechanisms cannot easily establish cooperative organisation. If they could, evolution would still have progressed toward increased cooperation between living processes, but it would have done so very quickly.

But in Chapter 5 we found that evolution could exploit the benefits of cooperation where it was able to establish a certain type of complex organisation. We saw that cooperation could evolve in organisations that include a manager who controls the other members of the organisation. Appropriate management could ensure that individuals would benefit in evolutionary terms from their cooperative effects on others, and would be disadvantaged by any harmful effects on others. Management could therefore align the adaptive interests of individuals and the organisation, ensuring that it paid for individuals to adapt cooperatively.

In Chapter 6 and in this Chapter we have seen that evolution will tend to produce managers whose interests are aligned with the interests of the group of individuals they manage. Managers will therefore do better in evolutionary terms when they use their power to support cooperation amongst the members of the group.

Putting these elements together, we can see that wherever life emerges in the universe, the immense potential benefits of cooperation will progressively drive the evolution of cooperative organisations of increasing scale in space and time. Life will begin when large molecules arise that are able to manage atoms and smaller molecules to produce new instances of the large molecules. If these larger molecules cooperate together, they can form organisations that will be even more successful in evolutionary terms. But cooperative organisations of molecules will not evolve unless it is in the evolutionary interests of the larger molecules to cooperate. A powerful manager can make it in the interests of a group of larger molecules to cooperate. The manager can use its power to organise cooperation.

Once a molecule that can act as a manager emerges, evolution will produce cooperative organisations of molecular processes. But the organisations will be small in scale. The scale of organisation that a particular manager is capable of controlling will always be limited. No manager can be all-powerful. Many organisations of limited scale will be formed. Within each managed organisation of molecular processes, the benefits of cooperation will be exploited. But there will be competition between organisations. The potential benefits of cooperation between the members of different organisations will continue to be unexploited. The barrier to the evolution of cooperation will have been overcome within organisations, but not between organisations.

This unexhausted potential for cooperation between organisations (and between the individuals in different organisations) will drive continued progressive evolution toward increasing cooperation. When powerful organisations emerge that can control other organisations, evolution will favour the formation of groups of organisations, each managed by a powerful organisation. The result will be cooperative organisations of organisations. But these larger organisations will still be of limited scale, the potential benefits of cooperation between the larger organisations will not have been exploited, and this unexhausted potential will drive a further repetition of this evolutionary sequence.

Continued repetitions of this process will progressively produce cooperative organisations of wider and wider scale. The potential benefits of cooperation will not be exhausted until all living processes in the universe are united in a single organisation of the largest possible scale. All the matter, energy and living processes of the universe will be managed into a single cooperative organisation. As we shall see in detail in Part 4 of this book, evolution on earth to date has organised molecular processes into small-scale prokaryote cells, prokaryote cells into larger-scale eukaryote cells, eukaryote cells into multicellular organisms, and organisms into societies. It is about to produce a unified cooperative organisation of living processes on the scale of the planet, managed by humans.

So the evolution of life has all the key features of a process that is fundamentally progressive: there are virtually unlimited potentials for improvement through the establishment of wider-scale cooperation between living processes. And these potentials can be exploited only through a step-by-step process in which each step necessarily builds on and improves on the previous step.

But there is another aspect of the progressive evolution toward increasing cooperation that we have not yet considered. We have seen that the formation of organisations in which a manager controls the other members of the organisation will enable cooperation to be exploited within the organisation. The alignment of the interests of the manager and the interests of the organisation as a whole ensure that the pursuit by managers of their own adaptive interests also serves the adaptive interests of the organisation. However, the extent to which the manager and other members of the organisation are actually successful in discovering the forms of cooperation that are best for the organisation depends on their adaptive ability. It depends on how good they are at evolving, how smart they are at searching out the best cooperative arrangements, and how innovative and creative they are at discovering new cooperative adaptations and adapting these to changing conditions.

Their evolvability will also determine how effective they are at discovering how to form cooperative organisations of wider and wider scale. Living processes that evolve by the genetic evolutionary mechanism will be slow to discover how to form cooperative organisations. The genetic mechanism has no foresight or ability to anticipate future evolution. It has no capacity to understand the progressive evolutionary sequence of increasing cooperation that I have outlined, and no ability to use the sequence to guide its evolution. The genetic mechanism will still exploit the potential benefits of cooperation, but only when it blindly stumbles on cooperation that proves successful. In contrast, when living processes arise that develop an understanding of the progressive nature of evolution, evolution can move far more rapidly and directly to produce cooperative organisations of increasing scale. Once organisms know where evolution is headed and develop the ability to use this knowledge to guide their own evolution, progressive evolution will accelerate.

In Part 3 of the book (Chapters 8 to 12 inclusive) we will look at how evolution has progressively improved the adaptive ability and evolvability of living processes on this planet. We will see where humans fit into this evolutionary sequence. Then we will look at what improvements in our adaptive ability are likely to be necessary for continued evolutionary success in the future. We will identify the adaptive abilities we will need as individuals and the abilities our societies will need if we are to contribute significantly to the future evolution of life in the universe. We will see that a significant improvement in our evolvability will come once we can use our knowledge of the direction of evolution to build cooperative human organisations of greater and greater scale.

PART 3

The Evolution of Evolution

8

Smarter Cooperation

A manager that is in full control of an organisation has the potential to establish a wide range of cooperative activities. In principle, such a manager is capable of putting together any pattern of cooperative specialisation and division of labour. But what particular cooperative activities should it organise? Of all the alternatives the manager could support, how does it discover those that are best for itself and the organisation? Unless the manager is able to discover the most productive forms of cooperation, it will not be able to fully exploit the benefits of cooperation. It is not enough that a manager is able to organise cooperation. It must also be smart. It must be able to discover the cooperation that is best.

Early cells, the first multicellular organisms, and human tribes all had the potential to construct an enormous variety of cooperative relationships between their members. But they had to discover those that were best, and adapt them as circumstances changed. And this did not come easily. Millions of years passed before the cells within multicellular organisms discovered how to specialise and cooperate to form an effective eye. Many more millions of years passed before the cells discovered how to cooperate to produce the complex brain found in humans and other mammals. And it was a long time before humans discovered how to cooperate together to build nuclear power stations, and to send people to the moon.

Organisations that are superior at discovering new and better cooperative adaptations amongst their members will have an evolutionary advantage. Their greater evolvability will enable them to exploit more effectively the immense potential benefits of cooperation. In turn, these benefits will reward improvements in evolvability. So the potential benefits of cooperation drive more than just the evolution of managed organisations of greater and greater scale. The benefits will also drive the evolution of managed organisations that are better at evolving. As evolution unfolds, living processes will get better at evolving, smarter at searching out the best ways to cooperate, and more innovative and creative at adapting their cooperation as conditions change[1].

In the next four Chapters, we will look at how evolution has progressively improved the adaptability and evolvability of living processes on this planet, and how it will continue to do so in the future. We will see that the existence of this progressive sequence of improvements raises significant questions for each of us: as individuals, where are we located in the evolutionary sequence? How much room for improvement is there in our adaptability and evolvability? Will we need new psychological capacities if we are to evolve in whatever directions are necessary for us to contribute to the future evolution of life in the universe? Can our knowledge of the direction in which evolvability improves point to how we must evolve psychologically if we are to contribute to future evolution?

The existence of an evolutionary progression in evolvability also raises important questions for humanity as a whole: how much room for improvement is there in the evolvability of human society? How could the evolvability of human society be enhanced? How could we improve the ability of our systems of government to search out new and better ways to manage our societies? Multicellular organisms eventually evolved brains and nervous systems that are far superior at adapting and evolving than the individual cells that form them. Must human society do the same? Must our societies evolve supra-individual adaptive processes that will be far smarter than the humans they contain? If so, how can we construct these supra-individual adaptive processes?

We will see that an understanding of the direction in which

evolvability evolves is critically important for future human evolution. Such an understanding will enable us to locate ourselves within the evolutionary sequence and to see what improvements in our adaptive capacities will be necessary in the future. It will point to what we have to do to improve our evolvability, as individuals and collectively. Becoming aware of the direction in which evolvability improves is an important step in the evolution of improved evolvability[2].

We will begin by looking in detail at how evolvability has improved during the past evolution of life. But before we start to trace this evolutionary sequence, we need first to develop a thorough understanding of the basic process that is used by living things to evolve and discover new adaptations. This will help us to see how the simplest version of the process could be progressively improved by evolution.

The principle that underlies this basic adaptive process is surprisingly simple. Living processes search for better adaptation by trying out changes. These changes can be made within the organism during its life, or in its offspring. The usefulness of the changes is tested by evaluating their effects on the organism in which the change is made. So better adaptation is discovered by trying out changes and then testing their effects to see whether or not they improve adaptation. Trial-and-error is at the heart of the basic adaptive process.

Importantly, this 'change-and-test' process[3] can work well even if it does not use any intelligence to decide the changes that are to be tested. The basic process can discover complex adaptations even if the changes that are tested are chosen randomly. It is the testing that sorts through these changes and discovers any that are better. Even if the changes are made randomly, some may be improvements, and these will be discovered when the changes are put to the test. Without any knowledge or insight into what might improve adaptation, a change-and-test process can discover better adaptations.

A number of examples will show how powerful this simple change-and-test process can be, and how widespread it is in living organisms. We will see that it is not only the basis of evolutionary mechanisms that discover adaptations and pass them from generation to generation. It also underpins physiological and other non-evolutionary adaptive systems that adapt organisms or groups of organisms only during their life, and do not produce change across the generations.

The genetic evolutionary mechanism itself is one of the simplest examples. In genetic evolution, changes are generated when an organism produces offspring. The offspring generally vary in a small number of ways from their parents and from each other due to genetic changes. The genetic changes are tested by their effects on the offspring that carry them. If a change causes its carrier to do better, the carrier will produce more offspring, and the numbers of individuals who carry the change will increase in the population. Eventually, all members of the population will carry the change, and it will be established as an adaptation. The genetic evolutionary mechanism will have discovered a better adaptation. When the environment changes, a different change might then do better. Once the new change has spread throughout the population, the mechanism will have discovered a better adaptation to the new conditions.

For an example, consider a hypothetical population of snow hares. The genetic evolutionary mechanism will tend to produce hares that have the thickness and length of fur that is best for the environmental temperatures met by the population. It will do this by trying out offspring with a variety of types of fur. Those with fur that is best suited to the conditions will be the most competitive. But if temperatures change significantly, offspring with different fur will do better. The length and thickness of fur in the population will change as the conditions met by the population change. In this way, the genetic evolutionary mechanism will adapt the type of fur in the population to track changes in environmental temperatures.

In contrast to genetic adaptation, individual organisms adapt physiologically by trying out changes within their bodies during their life. The changes are tested on the basis of whether or not they produce a useful effect within the organism. For example, warm-blooded organisms use such a process to discover adaptations that keep their body temperatures constant despite changes in their environment[4]. The animal tries out changes that influence the amount of heat produced within the organism, and the rate at which this heat is lost to the environment. It might change its metabolic rate, its level of movement and other general activity, the amount of its food intake, its posture, its rate of panting and sweating, the amount of blood flowing to its extremities, and the extent to which its hair or feathers are raised and lowered. The animal does not know in advance what pattern of changes are needed to maintain its temperature at the best level in the face of change in its external environment. This pattern is discovered by trial-and-error, by testing changes against their effects on the animal's temperature. In this way, patterns of adaptive change are made within the animal during its life to track changes in external temperatures.

Some animals are also able to adapt to varying external temperatures by trying out changes in behaviours that can affect heat gain or loss. The animal may move in or out of the sun and change its skin colour in the search for changes that will help maintain the desired temperature. Humans may try out different types and amounts of clothing to adapt to changes in temperature. Many animals also use simple change-and-test mechanisms for other adaptive challenges. Most complex multicellular organisms are able to use change-and-test processes to adapt other aspects of their behaviour, and to discover new behaviours that are better at meeting their adaptive goals[5].

Societies of humans and other animals also adapt using variants of the basic change-and-test process. A colony of honeybees can maintain the temperature of its nursery around 34 degrees C (93 degrees F) despite large changes in temperature outside the hive. This is the best temperature for hatching eggs and rearing young. Bees can increase the temperature of the nursery by clustering more tightly around it, raising the rate at which they metabolise sugars, and by flexing their muscles more often. They can reduce the temperature by fanning their wings to improve the ventilation of the hive, and by increasing the amount of water that is evaporated within the hive. Nothing in the hive knows the particular pattern of these behaviours that will maintain the temperature of the nursery at the ideal. The pattern that will maintain the best temperature is discovered by testing changes in these behaviours against their effects on the temperature in the nursery[6].

Human economic markets are an example of a process within human societies that uses a basic change-and-test mechanism to achieve adaptation. For example, markets can adapt the level of production of particular goods to the needs and preferences of consumers[7]. If insufficient goods of a certain type are being produced, manufacturers who increase production will be rewarded with higher profitability. This mechanism will increase production even if individual manufacturers are completely unaware that there is an emerging shortage of the product, or have no idea what is causing it. It will work even if manufacturers use only simple trial-and-error to decide their level of production. In this way, an economic market can adapt the level of production of warm clothing to track changes in demand as environmental temperatures vary over a series of winters.

These examples can also help us to see how the basic change-and-test process can be combined with more complex arrangements to improve the ability of the process to discover adaptations. The basic process searches for adaptation by trying out changes. It can discover improvements even if the changes are chosen randomly. But the search will be more efficient if the changes are chosen so that they have a better than random chance of proving adaptive. Change-and-test processes will do better if the changes they try out are non-random and are instead targeted at the types of changes that are likely to prove adaptively useful.

So the genetic evolutionary mechanism will be more effective if the genetic changes that are tested are non-random, and instead are more likely to be adaptive. We will see in the next Chapter that evolution has indeed established genetic systems that produce targeted genetic changes. And the physiological systems that adapt warm-blooded animals to differing external temperatures do not test out random changes within the organism. In high temperatures they test out changes that will cool the animal and reduce its internal production of heat. In low temperatures they test changes targeted at doing the opposite. The same is the case for the changes that are tested within beehives in the search for adaptation to varying external temperatures. They still use trial-and-error to discover the best pattern of changes, but do so far more efficiently by targeting the changes.

In some cases the arrangements that target physiological changes are established by the genetic evolutionary mechanism. Genes that cause physiological systems to target changes will do better than alternative genes that do not. But targeting can also be established by learning within the organism. If an organism discovers by trial-and-error that a behaviour pays off in particular circumstances, it will do better if it can learn to immediately try out this behaviour whenever the circumstances arise again. Learning avoids the repetition of costly trial-and-error[8]. The behaviour of young animals generally includes a high level of trial-and-error until they learn to target their behaviour more accurately at their particular adaptive goals.

The change-and-test process can be targeted even more accurately if the organism is able to form mental representations or models of itself and of its environment, and is able to test possible changes against these models mentally, before trying them out in practice. Instead of using the change-and-test process to try out actual changes in the real world, possible adaptive changes are first tried out mentally[9].

On this planet, we humans have the most highly developed capacity to search for adaptations using mental processes. In some circumstances our modelling capacity is so effective that it can completely eliminate the need for external trial-and-error. When our mental model of a situation can accurately predict the consequences of our alternative acts, we can mentally design an action that will directly achieve our adaptive goal. We then simply implement the changes that we see will achieve our goal. No external change-and-test process is necessary.

When we set out to solve an adaptive problem, it is these mental modelling processes that we are conscious of using. When we plan how we our going to cook our evening meal, think about how to fix a car engine, or imagine what we might have to do to improve our career prospects, we are using mental representations to come up with behaviours that will achieve our adaptive goals.

In contrast, we are not conscious of the simpler change-and-test processes that are continually adapting our bodies and internal organs to variations in temperature and to changes in the availability and usage of food, water and oxygen. We continually experience our mental processes, but have no experience of the workings of these other adaptive mechanisms in our bodies. When we think of how we might solve an adaptive problem, we tend to think of the mental processes we could use.

As a result, we do not have a good mental feel for how change-and-test processes in other living processes successfully solve complex adaptive problems without the use of mental modelling. We find it hard to see how the genetic evolutionary mechanism, physiological systems in other animals, and the adaptive processes in economic markets and insect societies can discover and establish complex adaptation without using the mental processes we associate with intelligence.

This blind spot in our understanding of adaptive processes has been particularly limiting when we have set out to design and adapt our economic and other social systems. We have a tendency to think that these complex systems can be best designed and adapted by the use of human intelligence. We think that if we collect enough information about the system, we can understand it sufficiently to decide the course of action needed to produce the result we want. However, we can rarely have sufficient information about complex and rapidly changing systems to make them predictable enough for us to adapt them in this way. The experience of centrally-planned economies has made this increasingly clear in recent years. Attempts to use central planning to match production levels to the needs of consumers have been spectacularly unsuccessful.

The alternative is to produce economic and other social systems that include their own adaptive change-and-test processes. An example is the economic market that we briefly considered earlier. A market system uses change-and-test processes to adapt production to match the needs and preferences of consumers. Such a process will be more effective if it can take advantage of the mental capacities of the participants in the system to better target the changes that are tested. But ultimately it is the systemic change-and-test process that adapts the system. And such a process can work even if the mental modelling used by participants to target their behaviour is ineffective. We will return to these issues when we consider in detail the future evolution of human societies.

Armed with this broad understanding of the nature of the basic process that living things use to evolve and adapt, we will trace the evolution of the evolvability of living things on this planet. Over the next four Chapters we will see how the potential benefits of cooperation have driven an impressive sequence of improvements in these abilities, and how this progressive evolution can be expected to continue into the future.

We begin in Chapter 9 by looking at how living processes evolved before genetic systems emerged. We will see how autocatalytic sets could evolve without genes, and how their evolvability could improve. Then we will move on to consider the evolution of the genetic system itself. We will look in detail at the evolution of the evolvability of genetic systems. We will see how natural selection has improved the ability of genetic systems to discover adaptations. Genetic systems produce a pattern of mutations and other genetic changes that is far from random. The pattern is biased toward changes that are more likely to produce useful adaptation.

But genetic systems do not adapt organisms during their life. Genetic systems cannot discover adaptations by trying out genetic changes within individual organisms. New adaptive processes have had to evolve to adapt individuals. Our physiological, emotional and mental adaptive systems are examples of what evolution has produced to fill this vacuum. In Chapter 10 we trace in detail the progressive evolution of the internal mechanisms that adapt and evolve individual organisms during their life.

In humans, the internal adaptive processes are now evolutionary mechanisms in their own right—through language, discoveries made by our adaptive processes are passed and accumulated from generation to generation. In Chapters 11 and 12 we will look at how these internal adaptive processes are evolving in humans at present, and how they are likely to continue to evolve in the future. We will see how we must improve our evolvability by developing new psychological skills if we are to contribute to the future evolution of life in the universe.

9

Smarter Genes

How smart at evolving are autocatalytic sets? How good are they at discovering new adaptations and passing them on from generation to generation? Does evolution tune and hone the evolvability of autocatalytic sets?

We have seen that an autocatalytic set is a group of proteins in a watery environment that collectively reproduces itself. Each protein catalyses (manages) reactions that lead to the formation of other members of the set. Collectively this produces a proto metabolism in which other molecules (food) are managed by the proteins to reproduce the set.

But autocatalytic sets of proteins do not contain genes. How can sets evolve? How can the members of the set discover better ways to cooperate with the other members, and adapt their cooperation as the internal and external environment of the set changes? How can they do this in ways that will be passed on from generation to generation of sets, producing evolutionary adaptation?

Autocatalytic sets are able to reproduce in some circumstances. This is because they become more likely to break up into smaller sets as they increase in size. The new sets that are produced in this way will compete with each other for the food and other matter that they need for their survival and reproduction. If a change arises within a set, the changed set may prove to be more competitive than others. A changed set that does better will grow in size faster, reproduce more frequently, and tend to take over the population of sets. As a result, a change that makes a set more competitive will be established as an adaptation possessed by all members of the population. In this way, natural selection operating between sets will test the effectiveness of any change that arises within a set[1].

But evolution can occur in this way only if changes arise within sets, and if the changes can be passed on from generation to generation. How can changes of this type arise?

Perhaps the simplest way is if different parts of a particular set become physically separated, and if the parts happen by chance to contain different components. Any part which contains members that can reproduce collectively as an autocatalytic set will be able to survive and reproduce as a separate member of the population. If the ways in which a set differs from its parent make it more competitive, it can take over the population, producing a population of adapted sets[2].

It is conceivable that this simple change-and-test process could even evolve sets that are better at evolving. For example, consider what will happen if sets differ in their tendency to separate into parts. If a set has a tendency to separate into different parts too readily and too often, it risks breaking up effective arrangements and producing daughter sets that are not competitive. Alternatively, if a set never separates into parts that are different, it will not evolve. It risks being out-competed by those that do. Sets that balance these tendencies will be better at evolving, and will be favoured by evolution. All change-and-test mechanisms face this particular dilemma. They must strike a balance between preserving their accumulated discoveries by minimising changes, and boosting the search for new improvements by trying out changes more often. Finding the best balance between conservation and change is an old evolutionary problem.

Nevertheless this simple change-and-test process is very limited in its ability to explore the evolutionary potential of autocatalytic sets. It cannot produce a set that has new members that were not also members of the parental set. It is limited to trying out different combinations of the same members. It cannot discover a new protein that might contribute more to the efficient operation of the set than existing members.

However, there is another change-and-test process that enables new members to be tried out in an existing set. A new protein molecule may form by chance within a set. This might happen through the chance interaction of particular molecules that come together in the right positions and under the right conditions to form the protein molecule. Alternatively, a number of molecules of one or more new proteins might drift into the area occupied by the set. If these new proteins are then reproduced as part of the set, and if they improve its competitiveness, adaptive evolutionary change will have been achieved[3].

But the ability of this process to test out all the types of proteins that could possibly improve the competitiveness of the set is strictly limited. It can try out only those proteins that, when added to the set, will be reproduced as part of the set on an on-going basis. If a new protein arises in the set by chance, or if it drifts into the set, it will not survive for long unless its formation is catalysed by the set. For a new protein that is not already reproduced by the set, this will occur only in exceptional circumstances. It will occur only if the addition of the new protein to the set causes changes that result in the new protein being reproduced by the set. The catalytic activity of the new protein must set off changes in the set that eventually result in the formation of the new protein within the set[4].

This is a very restrictive condition. But only proteins that meet this narrow requirement can be tried out and established by this evolutionary mechanism. If a protein does not meet this condition, it cannot be discovered by the mechanism, no matter how much it might improve the competitiveness of the set. Of course, this is an instance of the barrier to the evolution of cooperation that we looked at earlier. The barrier limits the extent to which an evolutionary process is able to exploit the benefits of cooperative organisation. There is a close connection between this barrier and evolvability. All instances of the barrier can be seen as a limitation in the ability of the relevant evolutionary mechanism to discover useful cooperation. And the emergence of managed cooperative organisation and other arrangements that overcome the barrier can therefore be seen as improvements in evolvability[5].

The limited ability of autocatalytic sets to discover new proteins was overcome once RNA began to manage autocatalytic sets to form proto cells. The RNA had the ability to cause the formation of proteins that were not reproduced by the autocatalytic set itself. It could therefore try out proteins that could not be tried out by an autocatalytic set alone. RNA could discover and reproduce proteins that would improve the competitiveness of the proto cell, but would not be reproduced by the set in the absence of the RNA[6].

RNA was particularly suited to searching systematically through the great range of new possibilities that were opened up. Each RNA molecule includes a long sequence of four basic units. Each sequence of units will produce a different sequence of the basic units that make up proteins. So a different protein could be tried simply by a change in the sequence of the basic units in the RNA. Importantly, mutations that change these sequences do not alter the basic character of the RNA molecule, and do not interfere with its ability to reproduce. So there were no limits to the types of proteins that could be tried out by RNA management. Not only did RNA have the ability to produce new proteins, it also had the ability to try out changes easily and systematically in each of the proteins that it produced. The result was a significant advance in evolvability.

The great evolvability of RNA meant that there was advantage in RNA eventually taking over the production of all proteins within the proto cell. Its greater evolvability could be used to search systematically for adaptive improvements in each protein that it produced. But RNA had to proliferate within the protocell if it were to be able to take over the production and evolution of all proteins. This could not occur immediately. It could take place only once arrangements were developed to suppress destructive competition between the various RNA molecules that produced the different proteins. Until the suppression arrangements we discussed in Chapter 6 were in place, the proliferation of RNA in the cell would have produced only destructive competition.

But the suppression of competition between RNA molecules meant that no evolutionary change-and-test mechanism involving RNA could operate within the cell. If it were to operate, alternative RNA molecules would have to be produced and tested within the cell. RNA molecules would have to mutate, reproduce and compete within the cell. But the suppression arrangements would prevent this from occurring. Competition, and therefore evolution, was prevented within the cell. Instead, mutated RNA molecules could compete only through competition between the cells that contained them. If a mutated RNA molecule improved the competitive ability of the cell that contained it, the cell and the molecule could breed up and take over the population of cells. The change-and-test process operated only at the level of the cell. Changes arose in the RNA within cells, but were tested only through competition between cells.

In essence, this is the genetic evolutionary mechanism that produces evolution in all single celled and multicellular organisms. Some of the details have changed: in later cells, DNA took over from RNA as the ultimate level of management. And in multicellular organisms DNA is an internal distributed manager rather than an external manager as in cells. But the evolutionary mechanism is essentially the same in all these cases despite the differences: competition between mutations is suppressed within the organism during its life, and mutations are tested by competition between organisms[7].

In the remainder of this Chapter, we will take a close look at the evolvability of the genetic evolutionary mechanism. We will ask whether evolution has shaped the genetic systems of organisms to increase their ability to discover useful adaptations. Has evolution produced genetic systems that are smarter than if they used only random trial-and-error? A central issue here is whether genetic systems have evolved any capacity to anticipate the future. Do genetic systems target mutations at the types of environmental conditions that are likely to occur in the future? Does the pattern of genetic changes contain a higher proportion of changes that are more likely to match future evolutionary needs?

To see what such an ability might mean in more concrete terms, we return to our hypothetical example of a population of snow hares. The population faces a critical environmental challenge due to temperatures that fluctuate considerably across the generations. If the genetic system of the population could target its genetic changes, it would not produce changes randomly across all the genes of the organism. Instead, it would produce changes that were more likely to pay-off, given the types of environmental challenges faced by the population. It would target its genetic changes towards producing a variety of lengths and thickness of fur, increasing the chances that the population could adapt quickly to fluctuations in temperature. Changes would be made less often to genes in which any change was likely to be harmful, and less often to genes in which changes would not be relevant to likely environmental change.

A central dogma of evolutionary biology is that genetic mutation is blind to future evolutionary possibilities. On this view, genetic mutation is not targeted at all. Whether a genetic change proves to be useful in discovering a new adaptation or in adapting to environmental changes is completely a matter of chance. Mutation is random in relation to future adaptive possibilities[8].

But this dogma is not based on hard evidence. No biologist has ever gone out and collected the evidence that is needed to test the dogma. To do so thoroughly, a biologist would have to catalogue the genetic variation produced in a population, assess whether it arises randomly across all genes, and, if there is any bias, determine whether the bias is correlated with the adaptive challenges likely to be faced by the population.

Apart from the practical difficulties in gathering this evidence, doing so has not been given a high priority by biologists. Most have considered that there are strong theoretical reasons to accept the dogma that mutation is blind to adaptive challenges. First: there is no obvious mechanism that would enable genetic systems to target changes at future adaptive needs. How could the simple processes that produce mutations 'know' anything about future evolutionary possibilities? Second, even though greater evolvability would be in the long-term interests of a population, it would not evolve unless it was also to the advantage of the individuals within the species. Smarter populations or species would be able to out-compete other species by adapting first when the environment changes, or by discovering new and better adaptations. But it is not enough that evolvability is good for the species. Unless it also continually benefits the individual genes that produce evolvability, these genes will be out-competed within the population, and greater evolvability will not evolve[9].

The great problem for genes that improve evolvability is that evolvability might not produce benefits continuously. The population might not encounter significant environmental challenges for many generations. When it does, any individuals that carry genes for greater evolvability can do better in evolutionary terms. These individuals are more likely to produce offspring who are better adapted to the new environmental conditions. But until the environment changes, individuals who carry genes for greater evolvability will not gain any benefit. In fact, if the arrangements that increase evolvability are costly, the individual will be disadvantaged.

This argument does not apply only to genes that improve evolvability. It also applies to the genes that establish the genetic change-and-test process itself. Foremost amongst these are genes that cause or allow mutations to arise in other genes. These genes will tend to be out-competed during periods in which there is no advantage to evolving. The cost of producing genetic changes will be a burden when there are no significant environmental challenges or other circumstances that make it worthwhile to try out genetic changes. There will be no pay-off for the change-and-test process if a population is well adapted to its environment, and if the environmental conditions are stable and unchanging. An individual will do better in these circumstances if it produces only offspring that are faithful copies of itself. Its young will be well adapted, like itself. An individual that instead produces some offspring that are changed from itself is producing maladapted young. Any change will be for the worse. All changes will be harmful[10].

For example, consider a hypothetical population of snow hares that contains a gene that produces mutations in other genes. The mutations influence the length and thickness of fur. An individual that carries this mutator gene will tend to produce some offspring with different types of fur. The mutator can do well if environmental temperatures change significantly every generation or so. The chances are that one of the offspring will have fur that is better for the new temperature. If so, it will out-compete other members of the population, including those that produced only offspring that were faithful copies of themselves. The mutated gene and the mutator gene that produced it can take over the population[11].

But if the temperature does not change over many generations, and if the population has the right fur for this temperature, the mutator will disadvantage any individual who carries it. The mutator will cause the individual to produce some offspring with different types of fur. The individual will be out-competed by others who produce only offspring with fur that is right for the unchanging environmental temperatures. It will be disadvantaged compared with individuals that do not try out anything different through their offspring.

Many evolutionary theorists argue along these lines that natural selection will favour zero mutation rates in most circumstances. They argue that organisms often experience stable environments for long periods, and do not face significant environmental changes every generation or so. During periods of stability, organisms would do better if they did not produce mutant offspring. Organisms would therefore suppress mutation if this could be done efficiently, they argue. They acknowledge that genetic mutations are continually produced in organisms, but suggest this is only because the arrangements needed to copy genes without mutations are too costly for the organism[12].

On this view, genetic evolution would have ended if it had ever discovered a cheap way to stop the copying errors that produce mutations. The genetic change-and-test mechanism exists not because it enables new adaptation to be discovered, but because mutations are unavoidable in practice.

However, this position is not intuitively attractive, and many biologists have searched for alternative explanations of how genetic evolvability might evolve. One approach is to look for ways in which the living or non-living environment of the species may be continually changing. If a population continually encounters adaptive challenges every generation or so, genes can be favoured that cause the production of some offspring that are genetically different[13]. As we have seen, if the environmental temperatures met by a population of snow hares are continually varying from generation to generation, genes that cause individuals to produce some young with different types of fur could be very successful. They could do better than those that simply produce offspring with the same fur as the parents.

But biologists have had trouble showing that key features of the environment of organisms arc in fact changing continuously in this way, every generation or so. The non-living environments of most species appear to remain stable for long periods, and the organisms appear to undergo very little evolutionary change during these periods.

The best candidate for a source of continual adaptive challenge is not the physical environment of a species, but other living organisms, particularly parasites[14]. When a parasite discovers a better way to exploit a species, the species will benefit if it discovers a way to counter the change in the parasite. In turn, the parasite will benefit when it finds a way around the counter move, and so on, indefinitely. The result is an arms race of move and counter move that produces continual adaptive challenges for the parasite and the host species. Computer simulations have shown that these continual challenges can provide an evolutionary advantage for mechanisms that continually try out genetic changes in the search for new adaptations[15].

But this parasite theory is too narrow to explain most of the genetic changes that are continually produced by populations of organisms. The genetic changes produced generation after generation by organisms are not limited to changes that might be useful in combating parasites[16]. Populations of organisms are continually testing changes in all aspects of the organism. This has been shown conclusively by experiments that subject organisms to artificial selection. Whenever animal breeders have set out to see if they can change a feature of an animal through artificial selection, they have generally met with success. Whatever features they look at, they find that genetic changes are being generated as the animals reproduce, and that these changes enable the animals to evolve in response to selection[17].

Another problem with the parasite theory is its prediction that genetic changes would cease to be produced if ever the arms race stopped. It is only the continual adaptive challenges produced by the arms race that provide a profit for evolvability. The parasite mechanism can explain the continual production of genetic changes only if the arms race never stops. But this condition is highly implausible. There is no good reason to expect that all species that continually produce genetic variety are continually engaged in evolutionary arms race with parasites[18].

So are most of the genetic changes produced by populations of organisms harmful but unavoidable? Do populations produce genetic changes largely because they are too costly to eradicate, not because they enable the population to evolve? Or can natural selection favour genes that produce evolvability? Can genes that cause genetic changes become established in a population because they improve evolvability, even though the population may experience long periods of environmental stability?

Theoretical work begun by American evolutionist Egbert Leigh in the early 1970's suggests that they can[19]. He was able to show that in restricted circumstances, natural selection will favour genes that cause populations to continually try out mutations. Further work carried out by a number of theorists has extended his conclusions to a wider range of circumstances and genetic changes[20]. This work has also shown that natural selection will favour genes that target their genetic changes so that the changes are more likely to meet future adaptive needs.

This new theoretical approach acknowledges that if a mutator gene is to be able to benefit from discovering useful mutations when the environment changes, it must be able to somehow survive periods when the environment is stable. The theory shows how a mutator can do this. It demonstrates that a mutator can always survive periods of stability if the rate at which it causes mutations is low enough. And this is the case even though all mutations produced during periods of stability are harmful because they change an organism that is already closely adapted to its environment[21].

The new theory begins by noting that individuals which carry the mutator will not always be disadvantaged in evolutionary terms. An individual will not be disadvantaged unless the mutator causes the individual to produce an offspring that carries a harmful mutation. When it does, the copy of the mutator gene in that offspring will be removed from the population. But until it causes a harmful mutation, each copy of the mutator will do as well as any other gene. The rate at which a mutator gene dies out of a population depends on the rate at which it produces harmful mutations. The lower the mutation rate, the longer a mutator can survive without discovering useful mutations. A mutator gene will not die out of a population until all of its copies have produced harmful mutations. And if the mutation rate is low enough, there will always be some copies of the mutator left in the population at the end of a period of stability. No matter how long the period of stability, a mutator can survive if its mutation rate is sufficiently low[22].

If a mutator can survive periods of environmental stability in this way, it can spread throughout the population when environmental conditions change and the mutator produces a successful mutation. When a copy of the mutator causes an individual to produce a useful mutation that takes over the population, the useful mutation will re-establish the mutator throughout the population. Provided the mutator and the mutation are closely linked genetically, all individuals in the population will eventually contain a copy of both. The mutator will hitch hike on the back of any successful mutation it produces.

However, this does not solve the mutator's problem permanently. Every time that it re-establishes itself in the population by discovering a useful mutation, it then begins to die out again as it produces harmful mutations. To survive, it must again discover a useful mutation before it dies out. And to survive permanently in the population, the mutator must produce a useful mutation each and every time, before it dies out. Unless this condition is met indefinitely, a mutator will not survive continually in a population.

But this condition can be met often in a typical population of organisms. Over long time frames, a population will become increasingly maladapted as environmental changes accumulate. All organisms become increasingly maladapted if they remain the same as time passes. No matter how stable an organism's living and non-living environment, it will eventually accumulate significant change, and features of the organism that were once adapted will no longer be. So as time passes, the success rate of mutations will get better and better. The less an organism is adapted to its living and non-living environment, the greater the likelihood that a random change in the organism will produce an improvement. As environmental changes accumulate, it therefore becomes increasingly likely that even random mutations will be useful. Mutators with a mutation rate that is low enough will be able to survive until the likelihood that they produce a useful mutation becomes a certainty[23].

If a population is adapted to its environment, all mutations are likely to be harmful. Many will be lethal because they damage the effective functioning of the organism. The remainder might produce organisms that can function, but they will be less adapted than non-mutants. However, even in a relatively stable environment, as environmental changes inevitably mount over many generations, and as the changes increasingly disadapt the organism, the chance that a mutation will produce an improvement increases. Provided a mutator is able to survive long enough, it will become increasingly likely that its mutations will be useful to an increasingly maladapted population.

To illustrate how such a mechanism can operate, we will look again at our hypothetical example of an evolving population of snow hares. Imagine that the environmental temperatures met by the population usually vary little over a period of 10,000 years, but that over a time scale of 50,000 years, significant change is likely to have accumulated. A population that is very well adapted to the temperatures at the beginning of a 10,000-year period will therefore generally be adapted at the end of the period. Mutations will be harmful during this period if they change features of the snow hare that are suited to the prevailing temperatures. So a mutator that produces a high rate of mutations that change these features will soon die out of the population.

Compare this with a mutator that instead causes an individual to produce a mutated offspring only once every 20,000 generations, on average. If this mutator was common in the population at the beginning of the period and there is one generation per year, the mutator is highly likely to be still common at the end of even the 50,000-year period. But a snow hare population that was adapted to environmental temperatures at the beginning of the 50,000-year period will be maladapted if the snow hares are unchanged at the end of the period. And there will be a much higher probability that a random mutation will produce a snow hare that is better adapted to the new environmental temperatures. If the mutator produces such a change, it will spread throughout the population again.

So mutator genes can survive in a population if their mutation rate is tuned to the rate of relevant environmental change. The environment does not have to be continually changing every generation or so to favour the evolution of this evolvability. A gene that produces evolvability can be successful if it operates over a time scale on which the environment is changing continually. And this will hold true no matter how slow the rate of change.

Natural selection can be expected to tune mutation rates to balance two opposing tendencies[24]: first, a lower rate will improve the ability of a mutator to survive periods of little environmental change. Second, a higher rate will improve the ability of a mutator to discover useful mutations before another mutator does so. The higher the mutation rate caused by a mutator, the more likely it will discover the first useful mutation when the environment changes, provided it still exists within the population.

It can be expected that the optimum mutation rate will be different for different genes within the organism. The ideal mutation rate will be higher for genes that establish features of the organism that are affected by rapidly-changing environmental conditions. It will be much lower for genes that produce features that cope well with all but rare environmental events.

There is growing evidence from studies of bacteria that mutation rates do in fact vary for different genes, and that these differences are not random. Some genes are much more likely to mutate than others, and it is because there is evolutionary advantage in doing so. The higher rates have evolved because they improve evolvability.

For example, a particular group of genes in the parasitic Salmonella bacteria have been found to produce proteins on the surface of the bacterium. The organisms infected by these bacteria use the proteins to identify different types of bacteria. When the immune system of an organism has come up with a way of destroying a particular type of bacteria, it will be used against all bacteria with the surface protein common to that strain. But if the bacteria can change its surface protein, it can escape the defences of the host organism, until the host again locks onto its particular type of surface protein. Bacteria will do best if they can adapt to the continually-changing immune system of the host by continually searching for types of surface protein that are not recognised by the host organism. Researchers have found that genes that produce these proteins mutate far more rapidly than other bacterial genes that do not have to deal with aspects of the environment that are continually changing at such a high rate[25].

Natural selection will not only favour the production of mutation rates that vary across the genes of the organism. It will also favour the production of a targeted pattern of mutations that is more likely to include mutations that will meet future adaptive needs. A mutator will be more competitive if the mutations it produces include fewer that damage the individual, and more that change existing adaptations in ways that are likely to be useful as the environment changes. A mutator whose mutations are biased in this way is likely to do much better than a mutator whose mutations are random.

Mutators that are better at targeting their mutations will be discovered by the normal genetic change-and-test process. General mutation will produce a variety of mutators. Different mutators will produce mutations in different genes and in different parts of genes. Of these, the mutators whose pattern of mutations happens to do better at matching adaptive opportunities through time will out-compete those that are less effective[26].

For an illustration, we will revisit our hypothetical population of snow hares that experiences significant changes in environmental temperatures every few generations. We will consider what happens in the following circumstances: a particular mutator produces mutations randomly within a number of genes that are responsible for growing fur. Mutations in some parts of these genes change the length and thickness of the fur. When such a mutation produces a type of fur that adapts the hares to the prevailing temperature, the mutation and the mutator will spread through the population. But mutations in other parts of these genes are harmful. They produce hares with no fur, with damaged fur, or with only patches of fur. None of these mutations will ever be useful in the temperatures met by the population. So the mutator that produces mutations randomly across all parts of these genes will produce both useful and harmful mutations, depending in which parts of the genes the mutations arise. But despite producing some mutations that are always harmful, such a mutator can survive in the population, provided it is always able to eventually discover an adaptive mutation before it dies out.

But what will happen if a new mutator arises that has the same mutation rate, but produces only mutations that change the length and the thickness of the fur? The new mutator does not produce any mutations in the parts of the genes where mutations are always harmful. When the temperature changes, this new mutator will be more likely to find an adaptive mutation first, before the original mutator does so. The rate at which it produces useful mutations will be higher. When it is the first to produce an improvement, the new mutator will spread throughout the population. In contrast, the original mutator will have missed an opportunity to re-establish itself in the population. It will continue to produce mutations that are harmful, and if it continues to be beaten by the new mutator in the search for useful mutations, it will eventually die out. The mutator that is better at targeting its mutations will be favoured by natural selection. The smarter mutator will win.

In this way, the genetic evolutionary mechanism can be expected to improve its own evolvability. It will discover and establish mutators that target mutations at future adaptive needs. But this search process is very inefficient. It relies on costly trial-and-error to discover better mutators, and most of the alternative mutators that are tried will be inferior. Eventually, evolution might improve the ability of the genetic system to discover better mutators. The genetic system may learn to target mutations in mutators, increasing the chances that mutations will produce a better mutator. But again, it will take a lot of costly trial-and-error to develop this ability. There is obvious room for improvement here. A system that could target genetic changes without having to go through this expensive trial-and-error process would have a substantial evolutionary advantage. It would be clearly superior to the types of mutational systems we have considered to this point. The result would be a further significant advance in evolvability.

Evolution overcame this fundamental limitation in the evolvability of mutational systems when it discovered genetic recombination, which is part of the wider process of sexual reproduction. The evolution of recombination and sex arguably have been the most significant advance in the evolution of the evolvability of genetic systems. The great majority of complex single celled creatures and multicellular organisms now reproduce sexually and use recombination. We will see that the immense significance of recombination is that it is far more efficient and effective at producing genetic changes that are targeted at future adaptive needs. Sex and recombination are successful strategies for organisms because they make them smarter at evolving[27].

How does recombination do this? The genetic changes produced by recombination are better targeted because of the way in which the changes are generated. Recombination produces changes by putting together different combinations of existing genes. Unlike mutational systems, it does not produce changes in the genes themselves. As we have seen, most changes to genes that are already functioning well are likely to be harmful. They will disrupt the effective operation of the gene, and produce a damaged organism. Instead recombination generates changes by mixing existing genes together in different combinations. So the basic building blocks that it uses to produce new effects are genes that are tried and tested, and that already work well in the organism. It does not risk making changes to existing genes that are proven performers.

It works like this. The cells of most sexually-reproducing organisms contain two sets of genetic material. One set is inherited from each parent. Each set includes a number of chromosomes, and each chromosome is formed of a long string of genes. Because the organism has two sets of chromosomes, it has a pair of each different type of chromosome, one from each parent.

Each egg or sperm produced by the organism will have only one set of chromosomes. So when it fuses with another egg or sperm to produce a new organism, the new organism will have two sets of chromosomes. But the single set of chromosomes in each egg or sperm is not just a set selected from the chromosomes of the organism that produced the egg or sperm. Each chromosome in the set is new. It is a combination of the pair of chromosomes of its type in the organism. So each chromosome in the egg or sperm is a combination of the chromosomes inherited by the organism from its parents. Each is produced by a process called crossing over. Parts of the chromosome from one of the organism's parent are swapped for the same part of the matching chromosome from the other parent. This produces a new, full chromosome that is a combination of the chromosomes inherited from each of the organism's parents. The result is that the organism produces eggs or sperm that contain chromosomes with different combinations of genes to those in its own chromosomes or in its parents'. It is likely that the new combinations of genes put together in this way will produce offspring that differ from their parents. Natural selection will test which of the new combinations are better adapted.

But what makes the new combinations produced by this process particularly effective is not just that they contain genes that have proven to be effective in the past. It is not just that recombination produces genetic changes without risking the wrecking of existing genes by mutation. The great advantage of recombination is that it generates new combinations that are likely to be advantageous in the future. It tends to produce changes that are shown by past experience to be likely to be adaptive in the future[28].

To see how it does this, we first have to look at the type of genes that accumulate in the genetic material and that are available for recombination. We will again use the hypothetical example of a population of snow hares in an environment where average temperatures fluctuate significantly every few generations. We will first consider how the genes that affect the length and fitness of fur are likely to evolve over time if the population adapts through the production and testing of mutations, rather than by recombination. Then we will consider what sort of genetic changes would be produced when recombination creates different combinations of these accumulated genes.

Any complex process in an organism will be the result of the action of many genes. So there are likely to be a number of genes that are involved in the production of fur. There are also likely to be many possible changes in these genes that will have an effect on the length and thickness of fur. When environmental temperatures change, the first mutation that arises that changes the length and thickness of fur in a direction that matches the new temperature will be established in the population. Each time the temperature changes, up or down, new mutations that modify the fur for higher or lower temperatures will be established in the population. It is possible that in some cases adaptation will be produced by a mutation that undoes a mutation that was established in the past. But, until many mutated genes that modify the fur accumulate, it will be more likely that the change will be produced by a new mutation in one of the many genes that influence fur production.

As a result, the snow hare population will accumulate a number of mutated genes that have had the effect in the past of increasing or decreasing the length and the thickness of fur to match temperature changes. The extraordinary effectiveness of recombination is that it produces changes by putting together different combinations of these genes—different combinations of genes that each tend to change the length or thickness of fur. Rather than change fur randomly in directions that have not been adaptive in the past, all the changes tried out by recombination will tend to increase or decrease the length or thickness of fur. This will prove to be a very effective strategy for producing genetic changes if environmental temperatures continue to change in the future, as they have in the past. In contrast, the pattern of genetic changes produced by untargeted mutation would be far less efficient at adapting the population. Many random mutations would be lethal, or would produce changes that had nothing to do with changing the length and thickness of fur. So recombination is superior because it continually recreates combinations that have proved effective at some time in the past. And it tends to produce new combinations that change fur in ways that have proved adaptive in the past.

These principles can be expected to apply to any feature of an organism that must adapt genetically to some changing feature of the environment. The population will accumulate a number of genes that each will have changed the feature in a direction that produced adaptation at some time in the past. Recombination will produce changes by putting together different combinations of these genes. Each new combination is likely to change the feature in ways that have produced adaptation in the past. As a result, if future environmental changes are similar to past changes, recombination will be far more successful than mutation at targeting genetic changes at future adaptive needs. And the pattern of genetic changes that are tried out by a population using recombination will be far from random in relation to future adaptive needs[29].

The rate at which a population tries out new combinations of genes, and the content of the new combinations will be affected by a number of aspects of the recombination process. The rate and content will be influenced by: the frequency of crossing over between chromosomes when sperm and eggs are produced (if there is no swapping there will be no new combinations, and the chromosomes from the parents will be passed on to all offspring unchanged); the proportion of the chromosomes that are exchanged; and the location of different genes on the chromosomes (genes that are closely linked on the same chromosome are less likely to be separated and recombined by crossing over)[30].

Because these aspects of the recombination process are themselves controlled genetically, they are evolvable. So natural selection can tune them to optimise the rate and content of the genetic changes that are produced through time[31]. As a result, we can expect that the rate at which recombination produces changes in particular features of the organism will be tuned to the rate at which environmental changes affect the adaptedness of the organism. And we can expect that the content of the changes will be tuned so that changes are targeted more accurately at the types of adaptive opportunities that repeatedly confront the population.

Consistent with these expectations, the rate of recombination has been found to vary widely across different regions of the chromosomes of organisms that have been studied. And artificial selection in the laboratory has been able to readily change these rates of recombination[32]. But little work has yet been done on assessing whether these differences in recombination rates are adaptive as a result of their ability to control the rate of production of genetic changes.

There are other ways in which evolution can improve the evolvability of organisms by modifying the pattern of genetic changes that they produce. It can do this by changing other factors that influence the genetic composition of offspring. For example, evolution can program organisms to select mates that are likely to contain useful genes, or to mate with non-relatives so that they produce young that differ more from themselves.

In all these ways, evolution can improve the evolvability of cells and multicellular organisms by shaping the pattern of genetic changes that they produce through time. Selection can tune the rate at which genetic changes are produced, and target them at likely future adaptive needs.

But it is not only by shaping the pattern of genetic changes that evolvability can be improved. The evolvability of a population of organisms depends on their ability to produce offspring with changes that are more likely to meet adaptive needs. But whether offspring will be better adapted will not be determined solely by the nature of the genetic changes that the population generates. It will also depend on what effects these genetic changes have on the way the offspring develop and function. It will depend on the way in which the genetic changes interact with the existing features of the organism to produce actual change in the organism. And this will depend on the way the organism is structured and organised.

From this we can see that selection can shape the pattern of changes produced by a population in a number of ways: it can modify the pattern of genetic changes, or it can change the organisation of the organism, or it can do both. If evolution were unable for some reason to modify the pattern of genetic changes, it would still be able to shape the pattern of actual changes that were produced by genetic changes. It would do this by changing the organisation of the organism so as to modify the effect of the genetic changes[33].

An example will make this clearer. Complex multicellular organisms have physiological systems that enable them to adapt to changes in their internal and external environment. These systems adapt the organism to changes that would otherwise disrupt its efficient functioning. The physiological and other adaptive systems also enable the organism to adapt to internal and external changes that occur as it develops from an egg into a fully-grown organism. Again, in the absence of these adaptive systems, the changes could damage the organism, and disrupt its proper development.

These internal adaptive systems can also enable the developing organism to survive genetic changes that would otherwise be lethal to the organism that carried them. This enables an organism to make fewer costly errors in the search for genetic adaptation. For example, consider two organisms that produced exactly the same pattern of genetic changes in their offspring: the organism with the better internal adaptive systems would produce fewer offspring that die or malfunction as a result of the genetic changes. The organism would take fewer errors to discover a particular adaptation. It would be better at evolving[34].

In recent years, various theorists have pointed to a number of other ways in which an organism could be organised to improve its evolvability. For example, if each of an organism's internal functions are organised into a separate module or compartment within the organism, evolution can explore changes in a particular module or compartment without necessarily disrupting the functioning of the rest of the organism[35]. Redundancy of function can also contribute to greater evolvability: if more than one component of an organism can perform a particular function, evolution can explore changes in one of the components without disrupting that function. This is particularly significant in the organisation of the genetic material. If there are multiple copies of a particular gene, changes in one of the copies can be made without loss of function.

One final example: evolution could readily explore new structures in organisms that are constructed out of basic building blocks that can be combined together in different ways to produce a wide variety of structures[36]. Small changes in the type, numbers and placement of building blocks could produce a diversity of structures. The recombination process would particularly lend itself to producing these changes. There is growing evidence that many key features of organisms are, in fact, constructed in this way.

But there is a fundamental problem with arguing that selection for improved evolvability is responsible for significant features of the way organisms are organised. If these features provide a pay-off only while a population is adapting to environmental changes, how do they justify their cost during periods of environmental stability? As we have seen, arrangements that produce genetic changes can survive these periods be minimising their cost to the organism. If the rate of mutation or recombination is sufficiently low, they can survive until the environment changes and they can provide an adaptive benefit. But there is no equivalent way in which a significant feature of the organism itself can minimise its cost to the organism during periods when it does not provide any benefit.

Can this difficulty be overcome in cases where improved evolvability helps the species as a whole to adapt and survive? Species that are better at evolving because they have features that improve evolvability can be expected to out-compete other similar species. They are likely to spread and produce new species at the expense of less evolvable organisms. But there is a problem. The longer-term advantage to the species will not maintain the features during periods when they fail to provide enough benefits to cover their costs. During these periods, organisms that have invested heavily in these features will be disadvantaged compared with those that have not. So the features will be maintained within the species only if they have other effects that are favoured by evolution. If this is the case, they will not owe their existence within the species to their contribution to improved evolvability. And changes in the features that improve evolvability will not be favoured unless they also have other advantages. But features that enhance evolvability will contribute to the success of the species[37].

Of course, this fundamental problem does not arise if populations are continually adapting genetically to their living and non-living environment. If this is the case, there will be a continual pay-off for features and processes that contribute to this adaptation. However, as we saw earlier, the environments of most animals are usually not thought to be continually changing in any significant way. Many species exist without apparent change for long periods in environments that seem stable. But, as we shall see now, this does not rule out entirely the possibility that highly-evolvable organisms are continually and profitably adapting to small changes in their living and non-living environment.

The living and non-living environment of any organism will always be seen to be changing if it is looked at on a scale that is fine enough.

Examined closely enough, everything is continually changing, everything is in a state of flux. Temperatures, wind, humidity, and the intensity of sunlight change from hour to hour and day to day. The characteristics of food organisms, parasites, predators and other members of the species are also changing continually. So if a population of organisms were good enough at evolving, it might be able to find profitable ways to adapt to these continuous environmental changes.

This point is clearly demonstrated by complex multicellular organisms such as ourselves. Our heart rate, blood pressure, breathing, metabolic rate, and many other features of our bodies are being adapted continually to small-scale environmental changes. And the pay-off from this continual adaptation is apparently sufficient to justify the considerable investments made by our bodies in the systems that produce this adaptation.

As genetic evolvability improved throughout evolutionary history, populations can be expected to have got better and better at adapting profitably to finer and finer environmental changes. But are highly-evolved genetic systems that use recombination and sexual reproduction good enough at evolving to continually find profitable ways to adapt? Are they continually producing highly-targeted genetic changes that will adapt the organism to the small-scale environmental changes that all populations experience continually, no matter how stable their environment appears? If they are, organisational features that can improve evolvability might continually be able to provide a sufficient pay-off. They may be favoured continually by natural selection at the level of the individual within the population, and not contribute only to the success of the species.

I think that most populations of organisms are adapting continually in this way. If we could view a greatly sped up movie of a population of organisms over many generations, the population would appear to be adapting as continually and effectively as does a complex organism during its life. The population would appear alive in its own right, continually trying out highly-targeted changes, fluidly discovering new adaptations as conditions change, revising adaptations as necessary, and so on. But my view is little more than an intuitive guess at this time. Our current state of knowledge is a long way from enabling us to decide this issue conclusively.

*　　*　　*　　*　　*

In this Chapter we have looked at the ways in which evolution can be expected to have improved the evolvability of the genetic evolutionary mechanism. And we have seen that the genetic mechanism is much smarter than has commonly been thought by most evolutionists during this century. The pattern of genetic changes that are tried out by genetic systems, particularly those that utilise recombination, can be expected to be far smarter than random.

However, it is also easy to see that the evolvability of the genetic change-and-test mechanism is limited in a number of ways. As an evolutionary mechanism, it could be much improved. Its limitations leave substantial potential for further improvements in evolvability. And, as we shall see in the next two Chapters, this potential has progressively driven the evolution of new and better evolutionary mechanisms, and will continue to do so.

An obvious limitation of the genetic mechanism is that its capacity to learn is restricted. The targeting of genetic changes involves a form of learning. However, when organisms such as ourselves discover an adaptation that is effective in particular circumstances, we can learn to produce it again whenever those circumstances arise again. This is the ultimate in targeting. Trial-and-error is completely eliminated. But a genetic system cannot do this. It has no way of sensing the environment to distinguish one set of circumstances from another. So it cannot learn that particular adaptations are useful in particular environmental conditions. As a result, the genetic mechanism is unable to respond to specific environmental events by immediately producing a genetic adaptation that it has learnt is effective in those circumstances.

Nor can the genetic mechanism construct complex models of how its environment is likely to change through time, and use these to determine what genetic changes it should produce. Unlike us, it cannot use internal models to plan ahead, to test possible changes before they are tried out in practice, or to maintain adaptations that have no immediate benefits but will pay off in the future. A genetic system can miss a major beneficial adaptation by a single mutation, and never know. And it is unable to use a model of the direction of evolution to guide its search for better evolutionary adaptation.

But it was not these limitations of the genetic mechanism that immediately drove the progressive evolution of new adaptive mechanisms. The key limitation that did this was the inability of the genetic mechanism to adapt the organism during its life. As we have seen, a genetic change-and-test mechanism is unable to operate within the organism. This is because destructive competition must be suppressed within the genetic managers of single cells and the genetic managers of multicellular organisms. If genetic changes could arise and compete within the genetic managers during the life of the organism, destructive competition would undermine cooperation. Evolution therefore favoured arrangements that suppressed the possibility of this competition. But without competition between genetic alternatives, new and better adaptations could not be discovered within the organism. Genetic management could not provide a change-and-test process to adapt the organism during its life.

The inability of the genetic mechanism to adapt organisms continually during their life meant that there was enormous potential for the evolution of new mechanisms that could do this. This drove the evolution of new change-and-test mechanisms within the organism. Initially these were not evolutionary mechanisms. The adaptations that they produced were not passed from generation to generation. But, as we shall see, in humans these internal adaptive mechanisms have evolved into new evolutionary mechanisms that have overcome many of the limitations of the genetic mechanism. And we will see that they have to evolve further in the future to overcome other limitations and to improve the evolvability of humans as individuals and collectively.

10

Smarter Organisms

An organism that is able to search for adaptive improvements during its life has an enormous advantage. It is able to use its experiences to discover new, innovative adaptations and to modify them as circumstances change. For example, it can try out new ways to get more food or to improve its ability to avoid predators. And when external temperatures drop, it can try out changes within its body to discover how to maintain its internal temperature.

But organisms could not use the genetic change-and-test mechanism to search for adaptive improvements during their life. As we have seen, the genetic mechanism could not try out new genetic possibilities within the organism. The genetic mechanism was unable to exploit the enormous potential benefit of adapting the organism during its life. This meant that any new adaptive mechanisms that could fill this gap would be strongly favoured by evolution[1]. Such an adaptive mechanism would provide immense evolutionary advantages to organisms that possessed it. And any improvements in the new mechanism would also be favoured by evolution. As a result, evolution has established new internal mechanisms that adapt organisms during their life. Examples include our physiological and nervous systems. Evolution has also produced a long sequence of improvements in the ability of these mechanisms to discover adaptation. This sequence of improvements is continuing today.

Somewhat paradoxically, these new adaptive processes have been discovered and established by the genetic mechanism. Even though the genetic mechanism could not try out new possibilities within an organism during its life, it was able to establish new adaptive mechanisms within the organism that could. The genetic arrangements that produced these new adaptive mechanisms would be favoured by natural selection. Individuals that were better at adapting during their life would pass on more of their genes to the next generation. The genes that produced new adaptive mechanisms within individuals would therefore do better in evolutionary terms.

The simplest adaptive arrangement that the genetic mechanism could install in an organism is one that is completely hard wired. For example, the arrangement might pre-program the organism to change in a particular way when a specific event occurs. The change in the organism would be adaptive if it enabled the organism to deal more effectively with the event. We are hard wired with a number of these types of adaptive arrangements: we are pre-programmed to produce saliva when tasty food is put in our mouth; and we duck without thinking when a rock is thrown at our head. But adaptations that are fully hard wired do not include a change-and-test process. No aspect of the adaptation is discovered by making changes within the organism, and then selecting the change that produces the best result. Hard-wired adaptive mechanisms are fixed and inflexible during the life of the organism[2].

Hard-wired adaptations are flexible only across the generations. They are discovered and shaped by the genetic change-and-test mechanism. They can be improved and adapted by the genetic mechanism as circumstances change from generation to generation. But nothing new is discovered within the organism. The discovery incorporated in a particular hard-wired adaptation has been made over the generations by the genetic mechanism, and the results have been pre-programmed into the organism.

The limitations of hard-wired adaptive mechanisms are obvious. Every part of them has to be discovered and adapted over the generations by the genetic mechanism. No use whatsoever is made of the experiences of the organism during its life. Potentially an organism can gain an enormous amount of knowledge during its life about what works and what does not. But an organism that adapts only in ways that are pre-programmed makes no use of this potential. If its living or non-living environment changes in ways that make a hard-wired adaptation ineffective, the organism can't try out changes then and there. It and its descendants remain ineffective until they produce offspring with genetic changes that improve the hard-wired adaptation for the new environmental conditions.

So the potential advantages of being able to search for adaptive improvements during the life of the organism drove the evolution of internal change-and-test processes. Evolution favoured genetic changes that established processes within the organism that could discover new adaptations during its life. These internal change-and-test processes discovered better adaptation in the same way that all such processes do: changes are made to the behaviour or to the internal functioning of the organism, and these are tested against their ability to improve the organism's effectiveness. For example, a change-and-test process might try out changes in the metabolic rate of the organism, the blood pressure, the amount of blood shunted to the limbs and muscles, the amount of time the organism spends feeding, its hunting techniques, its fighting strategies, or how it avoids predators. And these changes would then be tested within the organism against the results that they produce.

The first change-and-test processes established by the genetic mechanism can be expected to have been the simplest. Smarter processes that needed more complex arrangements would take longer to discover and establish. Countless generations of search by trial-and-error were needed before the genetic mechanism discovered and established the first complex brain. We can use this principle that simpler processes often evolve first to reconstruct the long sequence of improvements in adaptability that have occurred during the evolution of life on earth. We will start with the simplest form of internal change-and-test process, identify its limitations, consider how these might be overcome by slightly more complex processes, identify the limitations of these new processes, consider how they might be improved, and so on.

A change-and-test process, no matter how simple, must include arrangements that try out changes within the organism. It must also include arrangements that test the changes, selecting those that are best. In more complex change-and-test processes, both the pattern of changes and the testing arrangements would be able to be improved by learning during the life of the organism. Both would include their own change-and-test processes that would enable the organism to discover better ways to target the changes and better ways to test them. But in the simplest change-and-test mechanisms, both the process that produces changes and the testing arrangements would be hard wired into the organism. The simplest mechanism would try out a fixed pattern of changes, and test them against some fixed internal standard. The arrangements that target and test changes would be established and adapted by the genetic mechanism.

But before we consider how this simple type of change-and-test process might be improved, we need to understand more about the internal testing arrangements. What sort of internal mechanism could evaluate the changes made within the organism? What mechanism could tell whether a particular change is good or bad for the organism in evolutionary terms? How could the organism know whether it is better to increase or decrease the metabolic rate, raise or lower the blood pressure, direct more or less blood to the limbs, or to fight or avoid another animal that is competing with the organism for food?

Natural selection will tend to establish internal testing arrangements that are able to identify the changes that will improve the evolutionary prospects of the organism. To be favoured by evolution, the genes that produce testing arrangements must improve the evolutionary competitiveness of the organism. To do this, the genes must produce testing arrangements that evaluate the ability of changes to improve the evolutionary success of the organism. The testing arrangements that are best at doing this will do better in evolutionary terms. As a result, internal testing arrangements are tuned by natural selection to use testing criteria that are correlated with evolutionary success. Natural selection establishes testing criteria that are good indicators of evolutionary success. Testing arrangements are tuned so that an internal change that would contribute to the evolutionary success of the organism will also do well against the test criteria.

What sort of arrangements could do this? What sort of testing criteria would be correlated with evolutionary success? What test could you apply to an internal change that could indicate the evolutionary impact of the change?

Probably the simplest way to test changes is to see whether they can return the organism to an efficient state after an environmental event has dis-adapted the organism. If it is profitable for an organism to adapt to an event, the event will probably have some adverse impact within the organism. So changes could be tested against their ability to reverse the adverse internal impact of the event. For example, consider a fall in external temperatures that reduces an organism's temperature below the level that is best for its metabolism. To discover how to adapt to the falling temperatures, the organism could test internal changes against their ability to move the internal temperature back to the level that is best for the organism. As a further example, consider an organism that is chasing its prey. It will use up oxygen in its muscles, reducing the level of oxygen below the concentration that is best for muscles to function efficiently. Possible adaptive changes could be tested against their ability to restore the level of oxygen to the concentration that is best.

In both these examples, when a key aspect of the organism moves away from its most efficient state, changes are triggered that are then tested against their ability to restore the aspect to its ideal state. Such a change-and-test process can be described as goal directed. It has as its goal the maintenance of a key aspect of the organism in a state that is best for the efficient operation of the organism. The change-and-test process will search for a pattern of internal changes that will maintain this key aspect of the organism at the best level through time, despite changes in internal and external conditions[3].

The genetic evolutionary mechanism will tend to establish these simple change-and-test adaptive processes for whatever aspects of the organism are best kept constant. These aspects of the organism have been called essential variables because their maintenance in a certain range is essential for the efficient operation of the organism. It is worth using the resources of the organism to defend essential variables against disturbance[4].

The genetic evolutionary mechanism will tune each of these simple change-and-test arrangements so that the changes it makes are targeted at the particular essential variable that the process seeks to maintain. Evolution will favour a change-and-test process if the changes it tries out are more likely to achieve the goal of restoring the variable. It will also be favoured if it tries out changes only when they are needed, and if its changes are the cheapest way of achieving adaptation.

These simple change-and-test processes are still used widely within single celled and multicellular organisms to adapt their internal arrangements to changes in environmental conditions. The complex physiological systems that are continually adapting our bodies as internal and external conditions change are based on these processes. They make sure that our cells and organs get enough food and oxygen, operate at a rate that matches the needs of the organism, do not accumulate damaging levels of toxins, get rid of wastes, and operate at the best temperature[5].

But simple change-and-test processes are able to discover the most effective adaptations only for a limited range of adaptive challenges. In their search for the best adaptations they cannot use information about circumstances outside the organism, or about likely future events. This is because they use the actual immediate effects of events within the organism to target and to test the changes that they try out[6]. So if two events outside the organism have the same effects on the organism, a simple change-and-test process will try out the same changes and use the same test in the search for adaptation. It is completely blind to the cause of the events, and all it can do is respond to their effects within the organism. So it is unable to target behavioural changes at the particular type of outside event that has affected the organism. It is unable to assess the outside cause of an internal disturbance, and try out behavioural changes that are most likely to deal with the cause.

For example, the internal temperature of an organism might increase either because the general environmental temperature has gone up, or because a nearby object is on fire. An organism that adapts only through simple change-and-test processes will respond to the two events in the same way. In both cases it will search for internal changes that will reduce its temperature. Because both events have the same effect on an essential variable, the organism's response to each of them will be the same. It cannot target changes at the cause of the particular outside event that is disturbing the essential variable. To do so the organism would have to have sensory arrangements that could distinguish between the two different causes. And it would have to be able to use this discrimination to target the changes it tries out at the particular cause. Only then could it discover that if the increase in temperature is due to a nearby fire, it should move away, but if it is a general environmental change, it might be best to reduce its metabolic rate[7].

These simple change-and-test processes are also adaptively blind to future events and to the future effects of possible adaptations. They are only able to discover adaptations that immediately correct the disturbance of an essential variable. So a possible adaptation that has very useful future effects but that does not immediately restore an essential variable will not be discovered. No matter how valuable the future effects of a particular adaptation, a simple change-and-test process will not be able to discover it. It cannot take account of the future effects of the possible adaptations that it tests. It has no foresight or ability to anticipate.

For example, consider a predator that lies in wait for its prey at a water hole. This behaviour might eventually benefit the predator, producing a kill. But initially the behaviour will not restore any essential variable to its ideal range, so a simple change-and-test process will not discover it. For another example, consider an organism that has a spear thrown at it. Simple change-and-test processes will begin to adapt the organism only when the spear begins to enter the body of the organism. It is only then that the spear begins to disturb essential variables. This is indeed how a sea sponge would adapt to a spear thrown at it.

For these reasons, simple change-and-test processes in modern complex organisms are largely restricted to adapting the internal processes of organisms to actual disturbances in essential variables. They are of little use for discovering adaptations that intervene in events outside the organism or that produce future benefits. In the terminology used by the great English systems theorist Stafford Beer, these simple change-and-test processes adapt the organism for the inside/now. He contrasted them with the more complex adaptive processes that adapt the organism for the outside/future[8].

The inability of simple change-and-test processes to adapt the organism for the outside/future drove the progressive evolution of more complex change-and-test processes that could do so. Arrangements that could successfully exploit the potential benefits of discovering adaptations for the outside/future had an evolutionary advantage. The result has been a long sequence of improvements in adaptive ability that is still under way.

In order to develop a good understanding of where this progressive evolutionary sequence has been heading, we need to look at what an ideal adaptive mechanism would be able to do. This will give us an idea of the potential for improvement that existed in simple adaptive mechanisms, and the direction in which this potential would drive evolution.

If an adaptive mechanism is to evaluate possible adaptations properly, it must assess all the effects of the alternatives. In order to select the best adaptation, it must take account of all their effects, whether they are good or bad, or whether they arise within the organism, outside it, or in the future. Any limit to the ability of an organism to predict and take account of relevant events and effects that occur elsewhere or in the future will limit its ability to discover the best adaptations. Any relevant effect that is ignored can lead to the selection of inferior adaptations.

For example, if an antelope is unaware that a predator is lying in wait at a water hole, it will fail to adapt more effectively by moving to another hole. If a lion is unaware that a drought will soon mean that prey will be scarce, it cannot alter its priorities to build up more fat reserves as quickly as possible while conditions are good. And the ability of humans to adapt effectively is obviously dependent on how far we are able to look into the future to take into account the likely consequences of our acts. A person who takes into account only events a day ahead will adapt quite differently to a person who looks only a year ahead, and both will live a different life to a person who looks up to 20 years ahead. The first person would never plant a farm crop, the second would never do a university degree to improve career prospects, and even the third would not voluntarily pay into an aged pension fund for a large part of his working life.

We see most easily the limitations of a narrow ability to take account of the future effects of adaptations when we deal with organisms whose ability is narrower than ours. Dogs, cats and children appear particularly handicapped in their adaptive ability when we see them ignoring future dangers that we can easily foresee. Of course, our adaptive strategies might look equally as silly to an organism that could take account of the consequences of our acts over even wider scales of space and time than we can.

To meet the adaptive ideal, an organism or a society would have to be able to foresee all the relevant effects of its actions. What is relevant will differ depending on the scale of the organism or society. For example, if human society increases in scale and colonises other planets in the solar system and elsewhere, events and consequences over wider and wider scales would become relevant to the adaptation of the society. Any limit to its ability to take into account any of these relevant events would impair its adaptive ability.

We are currently a long way from this ideal. In part, this is because the ideal may never be able to be fully met. There may be absolute limits to the ability of an organism to predict the future consequences of its acts in a highly complex and dynamic environment. But in most areas we are obviously far from reaching these limits. There are many technological and scientific discoveries that are yet to be made. And there is also much for us to learn about wider-scale processes in the universe that will impact on the future of humanity. We are only just beginning to understand something of the large-scale progressive evolutionary processes that will determine our evolutionary future. Humanity has barely begun to accumulate the knowledge and abilities needed for it to adapt for its outside/future.

Although life on this planet has not yet reached this ideal, it has made considerable progress. The evolution of life on earth has seen a long sequence of improvements in the ability of organisms to take into account events outside the organism and in the future. The sequence began with simple change-and-test processes that were only able to take account of the effects of events within the organism itself. Since then, organisms have progressively evolved the ability to take account of the effects of their actions on events over wider and wider scales of space and time[9]. As these capacities have improved, organisms have used them to discover adaptations that are more effective when the more-detailed and wider-scale effects of the adaptations are taken into account. At each step in the progression, organisms have been able to take account of the effects of adaptation that they were previously blind to.

We will now look at a number of key milestones in this sequence of evolutionary improvements in adaptive ability.

The first major improvement in the ability to adapt for the outside/future required a capacity to sense the external environment. The organism had to have a sensory system that was capable of distinguishing between different circumstances in the outside environment. This enabled the organism to try out different behavioural changes in different environmental circumstances, and to discover that some behaviours restored an essential variable in one set of circumstances but not another.

Change-and-test processes aided by a good sensory system could take account of the different effects that possible adaptations might have in the outside environment. Simpler change-and-test processes used only information from within the organism to determine which possible adaptations were tried out. Only the impact within the organism of external events was used. As we have seen, these simple adaptive processes were blind to the nature of the particular external events that caused the internal disturbances. But with the development of sensory systems, a change-and-test process could also use information about the outside environment. For the first time, a change-and-test process could discover that it was best to try out different changes in different environmental circumstances, even though the internal disturbances were identical in each case.

To exploit fully the benefits of achieving this first milestone, organisms had to develop the ability to learn from their discoveries. The organism could discover by trial-and-error that a particular change would restore an essential variable in certain external circumstances. If the organism could learn from this experience, it would not have to repeat the costly trial-and-error search for adaptation whenever those external circumstances arose again. Instead, whenever the particular essential variable was disturbed in the future in the same external circumstances, the change-and-test process could then go straight to trying out the change that produced adaptation in the past. The organism would learn that a particular behaviour is likely to produce a desirable internal state in particular external circumstances. The organism could then apply this discovery to future adaptive challenges[10].

For example, an animal may learn that if it is cold during the day, it can increase its temperature by finding sunlight to rest in. But if it is cold at night, it may discover that the best way to increase its temperature is to curl up in the bottom of its burrow. And if an animal is hungry, it may learn that if it is standing on soft earth, digging is likely to produce food that will satisfy its hunger. But if the ground underneath its feet is rocky, it may discover that moving to another place is better than digging where it is. In both these examples, a change-and-test process that relied only upon information about the state of internal variables could not learn to target different behaviours at the different environmental circumstances. And without a capacity to learn, the animal would have to rediscover the adaptations by trial-and-error each time circumstances changed.

The effectiveness of these types of adaptive processes depend on their capacity to distinguish between different environmental conditions, to store the discoveries they make, and to use them in the search for adaptation in the future. Evolution has exploited the potential benefits of improved adaptability by enhancing these capacities in organisms. It has produced long sequences of improvements in sensory systems and in the size and complexity of the nervous systems that store and apply learnt behaviours[11]. This has progressively improved the ability of organisms to discover and learn behaviours that act on the outside environment to produce desirable internal states in the organism. The result has been the high level of ability to discover and learn behaviour that is found in rats, pigeons, and other complex multicellular organisms. In these large-brained species, individuals use change-and-test processes to discover and accumulate a wide range of useful behaviours throughout their life[12].

Humans are largely unaware of the functioning of the simple change-and-test mechanisms that adapt our internal processes. However, we are conscious of the operation of the more complex processes that adapt our behaviour to external circumstances. When our body detects that an essential variable that is maintained by our behaviour is outside its preferred range, we feel a need to take action to restore it. We feel motivated to search for behaviours that will do this. These alternative behaviours are tested against their ability to restore the essential variable. And when we find a behaviour that works, when we achieve the goal of restoring the essential variable, we are rewarded by feelings of satisfaction or pleasure, and by the ending of any discomfort.

So if we lack enough water in our bodies, we feel thirsty and are motivated to try behaviours that have got us water in the past in the type of circumstances we are in. When a behaviour succeeds in producing water for us, when it meets the test of restoring the essential variable, we are rewarded by the pleasure of drinking and by the satisfaction of our thirst. We are hard wired with a system that rewards us for behaving in ways that maintain our essential variables in preferred ranges. The result is that we tend to behave in ways that satisfy our immediate material needs.

But these more complex change-and-test processes are still fundamentally limited in their ability to adapt organisms for the outside/future. They are unable to search for and discover adaptations that produce only future benefits. They cannot take into account the future effects of possible adaptations. This is because they test possible adaptations only against their ability to meet the goal of restoring disturbed essential variables within the organism. Future beneficial effects have no immediate effect on current essential variables. Behaviour that has only future beneficial effects will not satisfy a current need for a higher internal temperature, or more water, or more food. These needs can lead only to the discovery of adaptations that produce immediate results.

To overcome this limitation, evolution had to produce a new motivation and reward system that was not based solely on maintaining essential variables. The genetic mechanism had to establish a new system that would test possible adaptations against their future benefits as well as against their immediate effects. The new system had to immediately reward behaviours that produced longer-term benefits, even though they might not deliver any actual immediate benefits[13]. If attainment of a longer-term goal meant that the organism had to achieve a particular immediate goal, the new system had to produce a need within the organism to achieve the immediate goal. The need would motivate the search for behaviour that could satisfy the immediate goal, and therefore produce the longer-term benefits. The new system would have to do this even though achievement of the immediate goal might produce no actual immediate benefit to the effective operation of the organism. Rather than test alternative behaviours against their immediate impact on essential variables, the new system had to test alternatives against their ability to produce immediate internal rewards that were proxies for longer-term benefits to the organism.

If reward systems of this kind were hard wired in the organism, the organism would be able to discover behaviours that have only future benefits. This is despite the fact that all the organism ever does is seek immediate reward by searching for adaptations to satisfy its immediate needs. The organism would be hard wired so that its pursuit of immediate rewards causes it to behave as if it takes into account the future benefits of its actions. The better a reward system was at producing immediate rewards for possible adaptations that have future benefits, and the better it got at making the reward proportional to the future benefits, the better the organism would do in evolutionary terms. Natural selection would tune these hard-wired arrangements to match the levels of immediate rewards to the likely future effects of actions.

Sexual activity is a particularly clear example of behaviour that is organised in this way. Behaviour that causes an organism to sexually reproduce does not produce any immediate beneficial effect on the functioning of the organism. It does not restore any 'natural' essential variable. Sexual reproduction provides evolutionary benefits to the genes that produce it, but only in the long term. For these reasons, in less complex organisms sexual activity is completely hard wired into the organism, as are other adaptations for the outside/future. There is no ability to adapt sexual behaviour during the life of the organism using change-and-test processes.

In more complex organisms, the establishment of an internal reward system for sexual reproduction enabled sexual behaviours to be adapted during the life of the organism. It also enabled sexual behaviour to be prioritised and integrated with other needs of the organism. The organism was no longer pre-programmed to act in a particular way when a reproductive opportunity presented itself. Instead, it was motivated to seek out reproductive opportunities, and to search for behaviours that would achieve successful sexual reproduction. The organism was rewarded psychologically with pleasurable feelings when it achieved sexual goals. And motivations for sexual activity competed with other motivations within the organism to prioritise the various behaviours of the organism. In this way, the reward system organised adaptive behaviours that had no immediate functional benefits.

Internal reward systems that took into account the future benefits of possible behaviours began to be used more extensively as multicellular organisms evolved complex social arrangements. This is because many of the benefits of social existence are not immediate. Many of the actions that animals must take if they are to live together harmoniously do not have any immediate beneficial impact on the operation of the organism. But the actions are in their long-term interests. For example, it might be in the long-term interests of an animal to submit to a dominant individual long before the dominant does it any physical harm. And it might be useful for an individual to react angrily to the actions of another that undermine its status in the group, even though the actions do not affect it materially straight away. As a final example, it may be in the longer-term interests of an individual to be motivated to care for others in the group, even though the individual will not benefit from this in any material way immediately.

In all these cases, if the individual is to adapt in ways that are best for its longer-term interests, it needs an internal reward system that immediately rewards the adaptive behaviour. The reward system must do this even though the behaviour does not immediately improve the operation of the organism. As a result, when multicellular organisms began to exploit the potential benefits of cooperative social organisation, evolution produced complex new internal reward systems. These systems could motivate and reward the search for social behaviours that provided no immediate benefit, but that served the longer-term interests of individuals by enabling them to interact more effectively with others in the group.

As multicellular organisms such as dogs, monkeys, elephants and apes began to form complex social organisations, the genetic mechanism expanded and diversified the internal reward systems into complex emotional systems. This produced social animals that experience a wide range of emotional feelings. These motivate, reward and punish behaviours that often do not immediately impact on the efficient functioning of the organism, but will in the longer term. We humans, the most social of multicellular organisms, experience a wide range of emotions and feelings such as fear, anger, guilt, love, frustration, curiosity, sexual pleasure, self-esteem, grief, delight, shame and depression. All these were initially tuned by natural selection to motivate and reward behaviours that would adapt the organism in ways that produced longer-term benefits.

For example, fear can motivate an individual to avoid future dangers, anger can motivate an individual to attack others to stop them from undermining the individual's status in the group, guilt and shame motivate adherence to group rules and norms, love can motivate an individual to care for others in the group, frustration can motivate more attention to problem solving, curiosity can motivate an individual to explore new possibilities, sexual pleasure motivates reproductive acts, a need for self-esteem can motivate the individual to improve its status in the group, grief can motivate an individual to take greater care of others in the group, and depression can motivate an individual to try a new way of life.

Organisms with complex emotional systems spend their lives searching for behaviours and ways of life that will produce desirable emotional states, and avoid unpleasant ones. If the genetic evolutionary mechanism has properly tuned the emotional system, the behaviours that produce these internal rewards will also serve the longer-term interests of the organism, and ultimately its evolutionary interests[14]. The social behaviour motivated by the emotional system will ultimately produce evolutionary success. Emotional systems and their goals are means to evolutionary ends.

But most organisms with complex emotional systems such as baboons, dogs, cats, horses and dolphins are largely unaware that their internal reward systems have been shaped by evolution for evolutionary objectives. To them, their internal emotional rewards are ends in themselves. They spend their lives in the pursuit of satisfying emotional states, oblivious that this is evolution's way of getting them to discover the behaviours that are best for producing evolutionary success. Evolution has produced in them a virtual reality that motivates and organises their behaviour. But for them it is their ultimate reality. They cannot see beyond it to its real purpose. Most humans currently fall into this category. They are unaware that the emotional goals that drive their pursuit of wealth, power, sex and social success are merely means to evolutionary ends, not ends in themselves.

Because these organisms are unaware of the ultimate goal and purpose of their emotional systems, they are unable to adapt and improve them during their life. They have nothing to judge the effectiveness of their emotional goals against. There is no process within the organism that can evaluate whether any changes made to emotional goals would advance the organism's evolutionary interests. If they tried out new emotional goals, they would have no way of assessing the longer-term effects of alternatives. They have no insight into the purposes of their emotional systems. If circumstances change, and the immediate goals established by the organism's existing internal reward system no longer produce evolutionary success, there is no adaptive process within the organism to change the immediate goals. The organism will continue to serve the pre-existing goals that are no longer effective.

In these organisms, only the genetic evolutionary mechanism has the capacity to shape and tune the goals established by the internal reward system. The genetic mechanism can do this by producing a variety of individuals that are hard wired with different emotional goals and motivations. Individuals with goals that are better at advancing the evolutionary interests of the individual will have more surviving offspring, and eventually take over the population. The change-and-test process that adapts and improves the goals and motivations established by the emotional system is the genetic evolutionary mechanism.

Emotional systems were a major step forward in the progressive evolution of improved adaptability. They enabled organisms to search for and discover adaptations that had beneficial future effects, but did not produce immediate benefits. Emotional systems enabled organisms to take into account the effects of their actions over much wider scales of space and time. They were a major advance over simpler adaptive processes that could take account only of the immediate impact of possible adaptations on the functioning of the organism.

But, as we have seen, even the most highly developed emotional systems found in the social mammals such as ourselves are limited in their ability to adapt during the life of the organism. The framework for the reward system is hard wired into the organism, and is limited in its flexibility during the life of the organism[15]. The immediate behavioural goals that are established by the reward system are not very adaptable. In contrast, the particular behaviours that the organism can use to obtain the rewards provided by its emotional system are not hard wired. The organism can adapt its behaviour to whatever is needed in specific situations. For example, most social mammals have emotional systems that reward behaviour that improves their social status and power. But they are free to search for whatever behaviour will serve these goals in the particular social circumstances in which they live. Organisms can change their behaviour in whatever ways are necessary to achieve their emotional goals, but they cannot change their goals. Their behavioural strategies are highly adaptable, but the goals set by their emotional system are not. Means are very flexible, but ends are not.

As a result, emotional systems have a limited ability to take advantage of the experience of the organism during its life. The emotional system has little capacity to use experience to improve its ability to take into account the future effects of behaviour. It is left largely to the genetic system to tune the reward system to take better account of future effects. If the reward system does not establish specific and immediate behavioural goals that adequately reflect the future consequences of behaviour, not much can be done about it during the life of the organism.

In the evolution of life on this planet, these limitations are being overcome by the development of a capacity to use mental models to guide the adaptive process. This capacity has evolved most fully amongst humans. With mental modelling, the organism is able to form a mental representation of how aspects of its environment will unfold over time. It can use these representations to see what effects possible adaptations will have in the future. In its most highly developed form, mental modelling can work out the consequences of a wide range of hypothetical behaviours in hypothetical environmental circumstances.

The evolution of mental modelling is a major improvement in the ability of organisms to adapt for the outside/future. It significantly boosts the adaptability of change-and-test processes[16]. It enables the future effects of behaviour to be taken into account in both the targeting and the testing of possible behavioural acts. Mental modelling can be used to mentally test possible adaptations before they are tried out in practice. This has a number of advantages: the organism can avoid behaviour that has dangerous future consequences; can work out which behaviour will produce immediate benefits without having to actually try out the alternatives; and can identify behaviours that are likely to pay off in the future.

Importantly, the models used by the organism can be improved continually as the animal accumulates knowledge and experience during its life. Where it discovers that a model does not accurately predict the future effects of its actions, the organism can revise the model to take account of its discovery. As the organism gains more knowledge of how its environment is structured, how the environment changes through time, and how different behaviours can produce different effects in different environmental circumstances, the ability of its models to target and test possible adaptation will improve. And this improvement in adaptability occurs during the life of the individual, without any involvement of the genetic evolutionary mechanism.

As the modelling capacity develops, individuals increasingly accumulate substantial stores of knowledge during their lifetime. This knowledge is extremely valuable to the individual, enabling it to use models to discover better adaptations by predicting the effects of alternative actions. For example, the more knowledge that an early human hunter accumulated about game animals, the better he would be at using mental models to predict which hunting strategies would be most effective. But the knowledge accumulated by an individual during his life died with the individual. So as a capacity for mental modelling evolved in early humans, there was an enormous potential evolutionary advantage to be had by any new arrangement that was able to transfer the store of knowledge between individuals. Any individual that could pass his accumulated knowledge to his offspring, or any individual that could obtain knowledge off others would be greatly advantaged in evolutionary terms.

Imitation enabled some transmission of adaptive behaviours between individuals. But it was only with the evolution of language that humans gained a comprehensive ability to share the knowledge they accumulated during their life. Through language, a discovery made by an individual could be passed to others and used by them in their modelling. Importantly, this enabled knowledge to be accumulated and built-on from generation to generation. Each new individual born into the population did not have to start again, with empty models. And in cooperative human societies, particular individuals could specialise in the collection of specific types of knowledge that were useful to the group. The result has been a complex and growing culture of adaptive knowledge that is passed from generation to generation[17].

As knowledge accumulated across the generations, the modelling capacity improved its ability to predict the consequences of possible behaviours. It became more accurate, could predict the consequences of wider ranges of events, and could evaluate the effects of possible adaptations over wider and wider scales of space and time. This improvement in adaptability continued the evolutionary progression that began with the emergence of the first simple change-and-test processes that adapted organisms for the inside/now. Since the first simple processes emerged, internal adaptive processes have progressively developed the capacity to take into account the effects of possible adaptations over wider scales of space and time[18]. The continuation of this progression is now enabling humans and humanity to build mental models of the formation and evolution of the universe.

Once knowledge could be accumulated across the generations, adaptive processes that used mental modelling became evolutionary mechanisms. They could discover adaptations and pass them on from generation to generation. Evolutionary discoveries could now be made throughout the lives of organisms. This was an immense improvement over the evolvability of the genetic evolutionary mechanism. As the modelling capacity improved, it began to make the genetic mechanism redundant. The genetic mechanism would take many generations to produce an adaptation to an environmental change. It would usually take many different genetic changes in many different individuals to find a better adaptation. But mental modelling could discover the same adaptation during the life of a single individual. Within a generation the discovery could spread to all members of the population. Modelling can operate far more quickly than the genetic mechanism. As knowledge accumulates and mental models improve, the genetic mechanism is increasingly pre-empted by mental modelling[19]. The genetic mechanism plays less and less of a role in adapting the organism.

But of even greater significance is the potential ability of mental modelling to take into account much wider consequences of possible adaptations than can the genetic mechanism. The genetic mechanism is largely limited to discovering adaptations that provide benefits during the life of the organism. If an adaptation pays off during an organism's life, it will enable an organism to pass more genes to the next generation. But this is not the case if the adaptation produces evolutionary benefits only for future generations, and none during the organism's life. The gene for such an adaptation is likely to die out before any longer-term benefits accrue.

In contrast, the modelling capacity enables an organism to take into account processes and events that unfold over longer time scales than its life. In particular, it can plan its adaptation in the light of the longer-term evolutionary trends and patterns discussed in this book. It can choose adaptive strategies that will contribute to the evolutionary success of its descendants and of its species. And the organism can support the formation of managed cooperative organisations which ensure that, as far as possible, it will capture the benefits of its support for future evolutionary success.

As we have noted, the only organism on this planet that has a well-developed capacity for mental modelling is humanity. We use this capacity to plan ahead, imagine alternative possibilities, invent and adapt technology, build structures such as houses and roads, radically modify our external environment for our adaptive goals, establish long-term objectives, imagine how we might change the world, develop strategic plans, design projects and successfully undertake activities that pay off only in the future, such as plant crops and feed animals. We undertake scientific and other research so that we can develop models that are more accurate, take into account the effects of our actions over wider scales, and predict the consequences of a wider range of possible acts and interventions in our environment. And we use language in its various forms to transmit knowledge between individuals so we can all use the best models to guide our adaptation. The result is an evolving culture of adaptive knowledge that grows from generation to generation and that enables humanity to progressively improve our evolvability by producing better models.

And we are on the threshold of developing the capacity to do something that no other organism on this planet has been able to do: to consciously use our modelling of the direction of evolution to increase our chances of participating successfully in future evolution. We are in the process of developing complex mental models of the direction of evolution. These will show what we will have to do individually and collectively to contribute to the future evolutionary success of humanity.

But being able to mentally model our evolutionary future does not mean that we will want to use these models to guide our behaviour. We will be like the band of hunter-gathers that we visited in our imagination in Chapter 1: they knew what they had to do for future evolutionary success, but did not want to do it. Their existing behavioural goals and predispositions clashed with what they would need to do to achieve future success. Simply knowing that they had to change their behaviour for future success did not make them want to do so. Likewise, rather than embrace evolutionary objectives, many of us will prefer to continue to pursue the values and goals established in us by past evolution. We will continue to use our energies to seek social status, self-esteem, power, wealth and the other goals that currently bring us emotional rewards.

The behaviours that will get us these emotional rewards will often clash with what we would have to do for future evolutionary success. As we have seen, our existing emotional reward system has been established by inferior and shortsighted evolutionary mechanisms. These mechanisms are blind to the long-term evolutionary consequences of our acts. They are unable to give us the emotional goals needed for future evolutionary success. If we are to be motivated to embrace evolutionary objectives, we will have to develop new psychological skills. We will need skills that enable us to free ourselves from the dictates of our pre-existing emotional reward system. We will need skills that enable us to find motivation and satisfaction in whatever forms of behaviour are needed for us to contribute to the successful evolution of life in the universe. Our task in the next two Chapters is to identify the new psychological skills we will need, and how they might be developed.

11

Smarter Humans

Organisms that can build complex mental models have the potential to model and understand the evolutionary processes that have formed them and that will determine their future. Once they accumulate sufficient knowledge about these evolutionary processes, they will be able to see their evolutionary future. Potentially, the organisms will be able to use this evolutionary knowledge to decide how they will adapt as individuals and collectively. The organisms will continue to test possible adaptations against their ability to satisfy shorter-term material and social needs. But they can also use their modelling ability to test adaptations against their evolutionary effects. This will enable them to choose adaptations that are also consistent with longer-term evolutionary success.

The development of such a capacity is a major step forward in evolvability. An organism that can see what is needed for future evolutionary success and can use this knowledge will be better at evolving. It will do better in evolutionary terms than an organism that is blind to the longer-term evolutionary consequences of its acts, and that is unable to consciously target its actions at future evolutionary success. Once life becomes conscious of its own evolution, its evolvability can improve significantly.

An organism that is unable to model evolutionary trends will be limited to optimising its behaviour only for the shorter-term effects of its actions. The organism's adaptive mechanisms will be able to discover behaviours that are successful only when the more immediate effects of the behaviour are assessed. The organism will be unable to take account of the longer-term consequences of actions that occur beyond the life span of the individual. As a result, the organism is likely to establish adaptations that are maladaptive when their longer-term effects are considered. And they will fail to discover adaptations that sacrifice shorter-term interests to achieve greater long-term evolutionary success. An organism that can take into account the likely future evolutionary consequences of its actions will achieve greater evolutionary success than one that must rely on the genetic evolutionary mechanism and other adaptive processes of more limited scope.

Humans are rapidly accumulating the knowledge to model the future evolutionary consequences of our behaviours. As outlined in earlier chapters, we are beginning to see the broader direction in which the evolution of life progresses. We are beginning to see what humanity must do if we are to participate in future evolutionary progress. Successful participation in the future evolution of life means we must continue to form cooperative organisations of larger and larger scale. To enable these organisations to fully explore the immense benefits of cooperation, they must be managed as far as possible so that individuals capture the effects of their actions on others, and therefore treat the other as self. And the organisations must be structured so as to maximise their evolvability. When humanity forms a unified, managed organisation on the scale of our planet, the planetary society must develop evolutionary mechanisms that enable the society to adapt not only for the inside/now, but also for the outside/future. A society whose adaptive ability is limited to tuning its economic and social systems for internal efficiency will not be successful in evolutionary terms. The society must also include mechanisms that enable it to adapt as a whole in relation to events outside the planet, whether the events arise from living or non-living sources.

Once we develop comprehensive models of larger-scale evolutionary processes and trends, we will see in detail what we must do if we want future evolutionary success. But how motivated will we be to want future evolutionary success for humanity? Will we want it enough? If future evolutionary success means behaving in ways that clash with our more immediate material and social needs, will we sacrifice these to continue to pursue future evolutionary success?

To be more specific, how ready are humans to put the interests of a planetary organisation ahead of those of our nation, ethnic community, and religion? Will we abandon any belief, prejudice and value that might stand in the way of our support for a truly global human society in which all individuals of all races and backgrounds are treated equally? What if this means a reduced standard of living? Will we accept and support the development of a capacity for the planetary society to adapt for the outside/future if this means a lower level of satisfaction of our immediate material, emotional and social needs?

The mere development of the capacity to use mental models to understand evolutionary trends will not make us want to adapt in the ways needed for future evolutionary success. By itself, it will not motivate us to put evolutionary success ahead of all our other adaptive goals and motivations. This is because our current goals and motivations are produced by our pre-existing internal adaptive processes, including by our emotional system. And these fail to take account of the longer-term evolutionary consequences of our actions. They have been established by shortsighted evolutionary mechanisms. Our existing adaptive processes have not evolved to reward and motivate the behaviours necessary for future evolutionary success.

When mental modelling is first developed, its main benefit is that it enables us to discover better ways to satisfy our pre-existing objectives and motivations. It helps us to find better means to our ends; it does not establish the ends themselves. Our modelling capacity has enabled us to develop sophisticated technologies and other ways of intervening in our environment. But these technologies serve our pre-existing goals and motivations. If we had different emotional goals and motivations, our technology would be different, as would our systems of government and other social arrangements.

The important point is that our existing objectives and motivations have been established by evolutionary mechanisms and adaptive processes that do not take into account evolutionary effects beyond the life of the organism. Obviously, these objectives clash with those that we would need if we were to pursue longer-term evolutionary success for humanity. So if we use our existing objectives and motivations to decide whether we want to pursue evolutionary objectives, we will decide against doing so. We will not feel motivated to do all that is necessary for future evolutionary success.

An organism whose motivations and objectives fail to take into account the evolutionary effects of its actions will not value objectives that do. Instead, the organism will use the immense adaptive capacity unleashed by mental modelling to get better and better at serving the goals of its existing internal reward systems[1]. It will not matter that these were established by limited and flawed evolutionary mechanisms and adaptive processes. When the organism discovers technological advances of great power such as genetic engineering and artificial intelligence, it will use them to serve motivations and objectives that ignore its longer-term evolutionary needs. The same will apply to the organisms' systems of government and other social arrangements. They will not serve the longer-term evolutionary interests of the organism. No matter how sophisticated its technology and social arrangements, the organism will not achieve evolutionary success. It will get better and better at achieving the wrong ends.

This evolutionary difficulty is likely to be struck by any organism that develops a comprehensive capacity for mental modelling. It will eventually be able to model and understand the larger-scale evolutionary trends that will determine its future evolutionary success. It will know what it has to do to continue to participate successfully in the evolution of life. But, initially, at least, it will not be motivated to do these things. Instead, it will continue to be motivated to use its growing knowledge of the effects of its actions to serve only its pre-existing objectives and motives.

But to continue to be successful in evolutionary terms, an organism must overcome this difficulty. To make a significant contribution to the evolution of life in the universe, an organism must be able to form highly cooperative and highly evolvable organisations on the scale of planets, solar systems, and galaxies. An organism will fail to be relevant to future evolution if it remains unorganised on a single planet, serving objectives and motivations established by flawed and shortsighted evolutionary mechanisms. We can be sure that the organisms that make a significant contribution to the future evolution of life in the universe will be those that develop a capacity to motivate themselves to pursue evolutionary objectives. Future evolution will belong to organisms that can free themselves from the goals and objectives of their biological and social past. The organisms that end up managing the matter, energy and living processes of large tracts of the universe will be those that develop the ability to adapt in whatever ways are required for evolutionary success, unrestricted by the goals and objectives implanted in them by earlier evolution. Those that do not develop this ability will be failed evolutionary experiments.

How can this evolutionary difficulty be overcome? Can evolution change the motivations and goals of an organism to align them with the dictates of future evolutionary success? Repeatedly during the past evolution of life on earth, the genetic evolutionary mechanism has significantly changed the internal adaptive goals of organisms. The cells that grouped together to form multicellular organisms had quite different adaptive goals to the solitary single-celled organisms from which they evolved. The amphibians that moved on to the land had very different internal goals to the fish that were their ancestors. And complex new internal reward systems were needed to motivate multicellular organisms to form social groups.

In all these cases, the new adaptive objectives were established by the genetic evolutionary mechanism. Genes that produced the goals that were consistent with the new form of life were favoured by natural selection. But the genetic mechanism is unable to overcome the specific evolutionary difficulty that confronts us. The genetic mechanism is largely limited to establishing features that pay-off during the life of the individual. Genetic evolution finds it very difficult to establish features that benefit only future generations. A gene that does not advantage the individual that carries it, but helps only future generations, will soon die out. This is true no matter how large the evolutionary benefits the gene delivers to future generations. The genetic mechanism cannot look far enough ahead to establish the motivations and objectives that are needed to produce longer-term evolutionary success.

Of course, this evolutionary difficulty could be overcome somewhat if the organisms were managed as members of a cooperative organisation. The management of such an organisation could redistribute benefits so that organisms that contribute to the future evolutionary success of the organisation are immediately supported and rewarded. This would align the shorter-term interests of the organisms with their longer-term evolutionary interests. But organisms that encounter this evolutionary difficulty will not want to organise themselves in this way. Organisms will invest in forming an organisation that produces longer-term evolutionary benefits only if this is consistent with their motivations and objectives. If they are not already motivated to pursue longer-term evolutionary goals, they will not invest in an organisation that is designed to assist them to do so.

The genetic evolutionary mechanism is unable to hard wire us now with the motivation and goals needed for longer-term evolutionary success. Evolution will not rewire the hardware of our brains and nervous system to realign our motivations and goals with longer-term evolutionary objectives. Does this mean that humanity has no alternative but to forever use its capacity for mental modelling to pursue our pre-existing motivations and goals, even though this would condemn us to future evolutionary irrelevance? Or is there another possibility? Can our motivations and goals change through modifications to our psychological software? Is it possible for humans to learn to reorganise ourselves psychologically so that we acquire the capacity to consciously choose our motivations, likes, dislikes, goals and objectives? Through our own psychological efforts can we learn the new ways of thinking and feeling that are needed for future evolutionary success?

To begin to answer these questions, we need to look more closely at the sort of psychological capacities that we would have to develop, and consider their feasibility. If we are to free ourselves from our biological and social past, we would have to develop the ability to find motivation and emotional satisfaction in whatever we have to do to pursue evolutionary objectives. This would mean, for example, that we would have to be able to drop our emotional attachments to any ideas, attitudes, beliefs, norms, values, religious systems and moral principles that were inconsistent with future evolutionary success. And we would have to be able to continually change our motivations and our likes and dislikes to fit in with whatever is required for future evolutionary success.

Ideally we would develop the ability to manage our motivational and emotional systems so that we could choose what it is that we want to do. Then whatever actions were demanded by evolutionary success, we could find them rewarding and satisfying. Once we could do this, we would be able to pursue future evolutionary success without sacrificing our shorter-term interests. And we would be able to transcend our biological and social past, able to modify the objectives and motivations established by this past where necessary, and able to choose new values and objectives consistent with evolutionary success. Each of us would become a self-evolving being.

This would be made easier if we could form human societies that redistribute benefits to reward behaviours that help achieve future evolutionary success. The ideal would be to align both our internal psychological reward systems and the external reward systems established by the society with evolutionary objectives.

Will humans change psychologically in the ways necessary to meet this ideal? Can we develop the psychological tools that would enable us to make this major improvement in evolvability? Will we be able to establish evolutionary success as an ultimate objective, and align all our pre-existing adaptive systems with this goal, ensuring that whatever behaviour will produce evolutionary success will be rewarded by our internal reward systems? Can we escape out biological and social past, and become true self-evolving beings?

In the remainder of this Chapter and in the next, I will demonstrate that as the modelling capacity of an organism improves, the organism will tend to undergo a sequence of psychological transformations. These changes move the organism toward making this great improvement in evolvability. Humans have been progressing through this sequence, and can be expected to continue to do so. Although there is no guarantee that humanity as a whole will finally takes this great step in evolvability, the changes currently under way are increasingly making it possible.

This approach immediately raises a key question: how will improvements in our modelling capacity help us develop the ability to align our motivation and reward systems with evolutionary objectives? Improvements in our ability to model accurately the effects of our actions on our external environment can obviously improve our capacity to discover effective adaptations. But how can it contribute to changing the objectives and values that these adaptations serve?

It can do this when the modelling capacity is turned inwards, and is used to model the individual's own internal adaptive processes. Our ability to mentally model our external environment has enabled us to manage the world outside us, and to intervene in it to achieve our objectives. In a similar way, the development of a capacity to form mental representations and models of our own internal adaptive systems can enable us to manage them and to modify their operation[2]. Once we become aware of our internal thoughts, motivations and emotional states, we can observe how they operate, what influences them, and what adaptive effects they have. We can use this knowledge to build mental models and representations of the operation of our mental, emotional and physical adaptive systems. As an individual learns to do this, the operation of these systems will increasingly become an object of consciousness. The individual can develop the ability to mentally stand outside his thoughts, motivations and emotional states, and gain freedom from them. Psychologically the individual separates into two parts—one that observes, and another that is observed. He no longer experiences himself as his thoughts or emotions. Increasingly he can stand back from them and mentally watch how they unfold, think about how they will operate and how effective they will be, and consider how they might be modified in the interests of other objectives or to improve their effectiveness. The individual's internal thoughts and emotions become like objects of consciousness in his external world—he will use mental models to discover how he can manage them for his own ends. Eventually, the observing part of the individual's psychology develops a comprehensive capacity to manage the other part.

This psychological capacity for self-management will be reinforced and strengthened as it discovers better ways to adapt the individual. It will find more and more ways to do this as the capacity for mental modelling improves. Mental modelling will increasingly improve at taking into account the effects of alternative behaviours, particularly the effects over longer time scales. As mental modelling accumulates knowledge, its ability to discover effective adaptations will increasingly surpass the ability of the pre-existing adaptive systems. Where our mental modelling sees opportunities to improve our adaptability, self-management will modify the operation of the pre-existing adaptive arrangements to implement the improvements. The adaptive benefits delivered by self-management will drive its improvement.

But competent self-management will not completely override and replace the pre-existing adaptive systems. Instead, it will modify them as far as is necessary to realign their goals with longer-term evolutionary objectives. Self-management operates like a new visionary Chief Executive Officer who takes over a corporation and implements for the first time a forward-thinking strategic plan. The elements of the strategic plan are designed to ensure that the company will be competitive and successful in the future despite changes in technology, markets and competitors. The plan models the future external environment of the company, and uses this to identify how the company should change the way it operates in order to continue to be successful.

The challenge for the CEO is to change the way the corporation operates so that it serves the new direction and goals. In most circumstances, he will be able to change the behaviour of the company to implement the plan without replacing or changing the nature of the employees. The employees will continue to have the same personal interests, objectives and internal motivations. But, if the strategic plan is implemented successfully, they will have to behave quite differently to before. Their behaviour will now also have to serve the ultimate objective of long-term success for the company. For this to be achieved, the interests of the employees will have to be aligned with the longer-term objectives of the company.

The CEO can achieve this by changes to the ways in which employees are managed, rather than by changes to their internal adaptive systems. The new management will create a new pattern of incentives and disincentives and other environmental conditions for employees.

These will ensure that employees find it personally rewarding to act in ways that serve the corporation's objectives. Within the internal environment of the corporation, when employees follow their personal interests, objectives and internal motivations, they will contribute to the effective operation of the company.

In the same way, successful self-management does not generally override, repress or abandon the pre-existing mental, emotional or physical adaptive systems that currently adapt us for shorter-term goals. The pre-existing systems are effective at adapting us to meet shorter-term requirements. So all that self-management needs to do is to modify the operation of the pre-existing systems only in so far as is necessary to take into account longer-term considerations. In this way it can align their operation with the longer-term goal of future evolutionary success.

We will now look in more detail at how improvements in modelling ability have driven improvements in the psychological evolvability of humans. In particular, we will look closely at how the continuation of these improvements tend to produce in humans a psychological capacity for self-management. When it is fully developed, this capacity would enable humans to manage their pre-existing adaptive systems so that they could find motivation and psychological satisfaction in pursuing longer-term evolutionary objectives for humanity, whatever this may entail. They would manage their internal emotional system so that it no longer rewarded behaviour that was inconsistent with future evolutionary success. Evolutionary self-management would allow humans to transcend their biological and social past so that they are free to adapt in whatever ways are necessary to achieve future evolutionary success.

It is useful to divide modelling ability into three broad levels[34]. The first, linear modelling, is limited in the complexity of the processes that it can model successfully. It cannot model a complex system in which a large number of components mutually interact. It can deal only with systems that can be analysed into components that interact in chains of causation that unfold step by step. It can understand systems that can be analysed by logical reasoning. So it is capable of modelling the outcome of the interaction of a small group of people over short time frames, the working of mechanical devices, and the movements of the planets about the sun.

In contrast, systemic modelling can successfully model the unfolding through time of complex systems with many interacting components. So it can model and understand a large social system, a flexible international corporation, or an ecological system. The third level, evolutionary modelling, is able to model the evolution of extremely complex systems over large scales of space and time. In particular, it can model the large-scale evolutionary processes that have formed us, and that will determine the future evolutionary success of humanity.

The progressive improvement of modelling capacity through these three levels is driven by the greater adaptive abilities that the improvements bring the individual. Individuals who improve their modelling will be better equipped to discover adaptations that satisfy their needs and goals, whatever these happen to be. As individuals proceed through the levels, they will be able to model the effects of their actions more accurately, understand the effects on their environment of a wider range of possible actions, and will be able to model the effects of their actions over wider and wider scales of space and time.

In part, these improvements in modelling capacity will result from the progressive accumulation by humanity of knowledge that is more detailed and that relates to processes that unfold over wider scales in space and time. And in part they will result from enhanced mental skills, and from technologies that aid human mental processes such as writing, diagrams, computer simulation, and artificial intelligence. At any point in human evolution, individuals will differ in their access to this knowledge, and in their ability to use it. So at any point in human evolution individuals will have different modelling capacities. And individuals may be able to apply a high level of modelling ability to some of their behaviours, but a lower level to others. The levels of modelling capacity are not mutually exclusive.

Individuals can use these improvements in modelling capacity to discover better ways to modify their external living and non-living environment. But as I have foreshadowed, individuals can also use the improvements to enhance the management of their internal adaptive processes. As the modelling capacity develops, it will be able to improve the pre-existing internal adaptive processes because it will be able to take account of effects that these processes are blind to. So, as we shall see in detail, each level of modelling not only corresponds to a particular level of ability to discover adaptations that impact directly on our external world. It also corresponds to a particular level of ability to self-manage.

We will now look in more detail at each of the three levels of modelling capacity. For each level, we will look first at how an organism can use it to mentally represent and understand the effects of the organism's interactions with its external environment. Then we will look at how an organism can use the modelling capacity to represent and manage its own adaptive processes. In the remainder of this Chapter, we will deal with linear modelling. Systemic and evolutionary modelling will be considered in detail in the next Chapter.

External Linear Modelling

The simplest form of modelling we will consider is linear modelling. It can be used by an organism to mentally model and predict the effects of its actions on its external environment. The ability of linear modelling to predict the effects of alternative behaviours will depend on the extent of knowledge used in the models. As organisms accumulate knowledge across the generations, they will be able to model with greater accuracy the effects across wider scales of space and time of a greater variety of possible actions. The actions that they model will eventually include the use of technology to produce specific adaptive effects on their environment. As knowledge accumulates, the organisms will become increasingly conscious of their external environment and how they can manipulate it to achieve their adaptive goals.

But linear modelling is limited in the complexity of external events and processes that it can model and understand successfully. It models how processes will unfold through time by following chains of cause and effect. It looks for causal relationships between events, and simulates what will happen in a particular set of circumstances by tracing how events will cause other events in a step-by-step manner. Linear modelling breaks processes down into parts, and looks for how simple step-by-step interactions between the parts can be used to predict how the process will unfold under different conditions. Analysis, reduction and logical deduction are its basic tools.

Linear modelling can work accurately and effectively for processes that are simple. But it is quickly overwhelmed as complexity increases. It is of limited use for understanding processes consisting of many interacting components that all contribute to how the process unfolds. These processes cannot usually be analysed into step-by-step chains of cause and effect. Linear modellers are unable to understand or predict the behaviour of these more complex processes. As a result, linear modelling is also limited in relation to the scales of space and time over which it can model the effects of possible adaptations. Over larger scales of space and time, most processes interact with other processes and become far too complex to understand by linear modelling. For these reasons, logical step-by-step analysis is notoriously ineffective for modelling and understanding the unfolding of complex systems such as human economic systems, ecosystems, human history and the mind and other complex adaptive systems.

External linear modelling is the mental capacity we use most often as we adapt consciously in our day-to-day life. It enables us to understand simple processes and to predict their future behaviour. We have built and designed our technology, our housing and our gadgets so that they can be readily understood and manipulated with linear modelling. The environment we have built around us is far more simple and mechanistic than our natural environment. It has been designed by linear modellers for linear modellers. In contrast, we need a capacity for systemic modelling if we are to manipulate our natural environment successfully.

Linear modelling is also the basis of most science as it is currently practised. The scientific method has been extraordinarily successful at understanding simple, linear processes and parts of more complex systems that can be reduced to simple processes. But the traditional scientific approaches have had little success in dealing with fields that cover complex systems such as ecology, psychology, economics, sociology and the other social sciences.

Humans that are capable only of linear modelling are unable to manage large complex organisations competently. In order to manage complex human social groups successfully, a manager such as a king or other ruler must be able to mentally model the effects on the group of alternative acts of governance. He (or his associates) must be able to mentally simulate the effects of his management on the group. For this reason, until at least some humans had moved beyond the limitations of linear modelling, humans were unable to form complex cooperative organisations managed by external managers. To be competent, external managers must be capable of some form of systemic modelling. It was not until about 10,000 years ago that human societies organised by rulers or other external managers were formed.

External linear modelling is used by the individual to form mental representations of himself, his physical, emotional and mental characteristics, and his skills. These mental representations are essential if the individual is to model effectively how he might interact with his external world to achieve adaptive goals. He must be able to include himself and his capabilities in his external models. But an individual who is limited to external linear modelling will not model how his mental and emotional characteristics might be changed. Without a capacity for internal modelling, he can model different ways of behaving, but not different emotional reactions and modes of thought. He will tend to treat his emotional responses, motivations and ways of thinking as fixed and given. They will not be objects of consciousness that the individual believes he can change.

The linear external modeller looks out at the external world and uses his models to search for ways of acting on the world to satisfy his adaptive goals. But in these models, the goals established by his internal reward systems are not treated as variable. Without a capacity for internal modelling, he has little ability to mentally model how his goals might be changed, and what effects these changes might have. He is largely unaware of the possibility of changing the key aspects of these adaptive processes, cannot consciously choose to modify them, and tends to take them as given.

But as the capacity for mental modelling improves, it inevitably will begin to clash with the organism's pre-existing adaptive systems. As organisms accumulate knowledge, their mental processes get better at modelling the external world. They get better at predicting how they can act on their environment to achieve their adaptive goals. Eventually, their mental knowledge will be superior in some cases to the knowledge embodied in their internal reward system and other adaptive processes. Their mental models will enable them to take into account effects of their actions that their pre-existing systems are blind to. They will see that acting in the way dictated by their emotional and motivation systems may be against their longer-term interests in some situations. For example, as we humans mature and learn more about the consequences of our acts, we become aware that in some circumstances it might be best not to respond to emotional impulses such as anger and sexual drives. We use mental modelling to see that acting immediately on these impulses may prevent us from gaining greater emotional satisfaction in the longer-term.

The external linear modeller will not be able to resolve easily these inevitable clashes between his mental modelling and his pre-existing adaptive processes. Earlier evolution will not have given his mental processes the capacity to manage his motivational and emotional systems. When the capacity for mental modelling first develops, it does not have the knowledge or ability to adapt the organism as competently as the pre-existing adaptive processes. To have given it the power to manage these processes before it was competent to do so would have been disastrous. Without the ability to manage his motivational systems, a linear modeller is unable to ensure that he will be motivated to implement any superior adaptations identified by his modelling. A linear modeller might be able to see longer-term advantages in particular actions, but will be unable to manage his internal adaptive processes so that he will find the actions emotionally satisfying. He will continue to be motivated and rewarded for the same actions as before. For example, we might see mentally that dieting is in our longer-term health interests. But this does not make fatty foods less tasty, or dieting pleasurable. And we might see that locking ourself in a room and studying will pay off eventually with a better job. But this does not make study satisfying, or our hobbies less enjoyable. We have little capacity to change these motivations and emotional responses to ensure that they support the findings of our mental modelling.

In order to do what is suggested by his mental modelling, the external linear modeller is likely to attempt to repress and override existing motivations and emotional feelings. This may enable him to improve on his pre-existing emotional responses in circumstances that are uncomplicated. It may pay to override emotional responses where the advantages are clear-cut. But the linear modeller is not well equipped to improve on his existing systems in more complex situations. The problem is that external linear modelling is not competent to assess the consequences of overriding emotional responses in most situations. It cannot model and therefore cannot understand the effects of choosing to ignore internal rewards or motivations. The linear modeller has no comprehensive understanding of why his emotional system provides rewards and motivations for particular behaviours but not others, and why it produces various emotional states in particular circumstances. The individual is not conscious of why his adaptive systems operate in the way they do. The individual is therefore in no position to use mental modelling to decide whether to override or repress particular motivations or internal rewards.

The problem is made worse because the individual does not know that his modelling capacity is limited in these ways. He is not conscious of the limitations of his consciousness. To know these, he would have to have modelled his modelling capacity. He would have to have a capacity for complex mental self-reflection. But he does not have this ability. The individual will be unaware that his mental awareness is grafted on top of a sophisticated and complex hierarchy of pre-existing adaptive processes that routinely and continually solve adaptive problems that he has not even begun to understand.

So an individual who uses only external modelling to determine how he adapts and behaves will have difficulty in integrating his conscious, mental adaptation with his pre-existing emotional and physical adaptive systems. He will continually make mental decisions that serve some internal goals but conflict with others. And the mental models he uses to make these decisions will not have the knowledge or ability to make them competently. He may try to use his mental modelling to decide whether or not to respond to particular emotions, motivations, and physical needs, but he will have little knowledge of why these exist or what specific adaptive functions they perform.

A typical example of the late 20th century is an individual who single-mindedly pursues career goals, repressing and ignoring the internal physical and emotional signals that indicate he should also serve other adaptive needs. In extreme cases, if the individual continues to ignore stress and depression, the result is physical, emotional and mental breakdown.

The problem will worsen as the capacity for external linear modelling grows. As knowledge accumulates, the growing ability of modelling to discover effective adaptations means that it will be used increasingly to determine how the individual behaves. More and more, mental modelling will guide the individual on how to achieve the adaptive goals established by his internal reward systems. The result will often be that the individual ignores and represses some of the motivations and needs associated with his pre-existing adaptive system, even though these may be essential for producing behaviour and adaptations that are critically important for the effective operation of the individual.

The need to integrate mental modelling with the pre-existing physical and emotional adaptive systems is a problem that will be encountered by any organisms that begin to develop a capacity for mental modelling. Wherever in the universe mental consciousness emerges, it will initially clash with the processes that have adapted the organism up until then. As the mental system develops, it will be used increasingly to control and run a complex, multi-level organisation that it does not understand or appreciate. It will have ultimate power over a complex system it is only dimly aware of. Initially it will have the power but not the wisdom to manage the organism, and it will not know itself well enough to see its own limitations. On every planet where mental consciousness emerges and develops, there will be a demand for personal growth programs that promote the development of psychological skills that can deal with this problem. The organisms that contribute most to the future evolution of life in the universe will be those that successfully overcome the problem by developing the higher levels of modelling capacity that we will discuss in the next Chapter.

The linear modeller is like the Chief Executive Officer of a large modern corporation who develops a comprehensive vision of what the corporation must do for future success, but has little understanding of the internal processes that adapt the corporation. He has control over the direction of the company, and knows where it should be headed. The CEO has good mental models of the external environment of the company, but lacks effective models of its internal workings. He has little understanding of the internal patterns of incentives and disincentives that the corporation creates for its employees, and how this impacts on their motivations and work performance. He has little knowledge of how employees will react to particular changes and management actions.

When he sets out to change the way the corporation operates, such a CEO is likely to fail to motivate his employees to go with his new vision. Announcing the changes will not be enough, because this will not change the pattern of incentives and disincentives created by the internal environment of the firm. It will not change the corporation's pre-existing internal reward and motivation systems. Employees will not change their behaviour if they are asked to do things that do not pay off within the corporation, and to stop doing things that are rewarded. For example, asking employees to take more risks will not produce much change if they have discovered it is best to avoid risk because mistakes reduce promotional opportunities. And if the CEO attempts to override these pre-existing reward systems coercively, he will produce the internal double binds, stresses and low morale that is common in poorly managed modern corporations.

If the CEO is to implement his new vision successfully, he and his executives have to know sufficient about the adaptive characteristics of the organisation to see how the pattern of incentives and disincentives should be changed to bring the employees along with his vision. The operation of the pre-existing internal reward systems of the corporation will have to be modified. Changes will have to be made that align the interests of all levels of the organisation with the future vision, but without disrupting functions and behaviours that will continue to be needed by the organisation. Unless the CEO develops the capacity to do this, he can have a vision and he can explain it to employees, but he will not be able to implement it effectively. Employees will continue to adapt to the incentives and disincentives that they continue to encounter. If a capacity to model the future is to be used to guide the adaptation of the corporation, it must be integrated with the pre-existing processes that adapt the corporation on a day-to-day basis. And to achieve this, the CEO must have effective models of the internal operation of the company. Without these models, he cannot manage the pre-existing processes to realign them with his vision.

The need to integrate the capacity for conscious mental modelling with the pre-existing adaptive systems of the organism can be expected to drive a long sequence of psychological evolution. In this sequence, the modelling capacity will progressively develop comprehensive models of the pre-existing adaptive processes and of their adaptive effects. It will develop the ability to manage them where there is advantage in doing so. Significant advantages can be expected where self-management is able to improve the pre-existing adaptive processes. At first, the modelling capacity will not have the knowledge or wisdom to do this. But as knowledge accumulates, the modelling capacity will increasingly be more effective at discovering better adaptations than the pre-existing processes. It will be more accurate, and be able to assess longer-term consequences. Psychological processes will have a considerable advantage if they can use this modelling capacity to manage and revise the pre-existing processes to produce better adaptation.

Linear Self-Management

The first major step in the evolution of self-management in humans began with the use of linear modelling to model internal adaptive processes. As the capacity developed, mental, emotional and physical adaptive processes became objects within mental models, and therefore objects of consciousness. Instead of the individual experiencing himself as his thoughts and emotional states, he could experience himself to some extent as being outside them, and able to treat them as objects of consciousness that could be influenced. His thoughts and emotional processes were no longer taken as entirely fixed and given.

But the limitations of linear modelling also restricted the level of self-management that it could support. Linear modelling can understand only simple processes that can be effectively represented by chains of cause and effect. It is unable to model complex systems, and is largely limited to modelling the effects of possible adaptations over relatively small scales of space and time.

As a result, linear modelling is able to model and understand only simple mental, emotional and physical adaptive processes. It can model only those that take into account effects that can be followed by linear modelling. If a particular adaptive process deals with circumstances that cannot be modelled and understood by linear modelling, linear modelling will be unable to competently model the consequences of modifying the adaptive process. It will not be able to understand why the adaptive process is structured the way it is, or to follow the effects of modifying it.

Linear self-management is particularly limited at understanding the emotional system. As we have seen, much of the emotional system was established by natural selection to provide internal rewards for behaviours that adapt us in our social life. The adaptive problems we encounter in our social interactions are often complex, and cannot be fully understood by linear modelling. The reasons why our emotional systems reward particular behaviours and not others are often outside the understanding of a consciousness that uses only linear modelling. And if linear self-management attempts to modify and improve upon these emotional processes, it will not do so effectively.

If linear self-management over reaches itself and attempts to manage tightly all of the organism's adaptive processes, it can produce a personality similar to the parts of the external environment that are designed and built by linear modellers. The personality will tend to be mechanistic, overly simplified, inflexible and rigid, and ultimately maladaptive.

For these reasons, a linear self-manager will be unable to fully integrate his mental modelling with his pre-existing emotional and physical systems. He will be unable to use his mental modelling to competently manage the pre-existing systems, and unable to resolve the conflicts that inevitably arise. As we will see in detail in the next Chapter, it is not until an individual can develop more complex capacities for self-management that this integration can be achieved.

However, linear modelling can be particularly effective at modelling linear mental processes. This is because these mental processes and the environmental circumstances that they model are simple enough to be dealt with adequately by linear modelling. Linear modelling has the potential to model itself. An individual who is capable of linear modelling has the potential to improve his adaptability by developing the capacity to manage his own mental adaptive processes.

Individuals capable of a high degree of linear self-management are able to mentally stand outside their thoughts, treating them as objects of consciousness that can be examined, analysed and influenced. They can accumulate knowledge about rules of logical inference and deduction, and can develop the ability to use these to test the validity of their thought processes, discarding those that do not meet the tests. They are able to critically evaluate their own thoughts and ideas.

Linear self-managers can develop the ability to learn how to learn. They can analyse the thought processes and mental strategies they use to solve problems, mentally model alternative strategies, and implement those that are shown by their modelling to be more effective. In this way they become increasingly conscious of their mental processes, and use the power of mental modelling to search for improvements in their mental processes.

The capacity to model their own mental processes may also enable linear self-managers to deal with thought processes that are maladaptive, such as unproductive worry. But this capacity can itself be used maladaptively. The individual may discover that he can manage his thoughts in a way that will produce desirable feelings and avoid unpleasant emotional states, even when he is poorly adapted to his external environment. For example, he may be able to think positively in the face of impending disaster, or learn how to treat the external world as an illusion, reducing its ability to produce undesirable emotional states. This maladaptive self-management short circuits the ability of the internal reward system to motivate behaviour that enables the individual to function more effectively in its external environment.

As the capacity for internal linear modelling develops, it will also undermine a number of the adaptations that organise cooperative social behaviour amongst humans. We saw in Chapter 7 that distributed internal management can organise the members of a band or tribe to behave cooperatively. This management consists of a set of norms and inculcated behaviours that are reproduced in each member of the group. The inculcated behaviours predispose individuals to cooperate in situations where it otherwise would not be in their individual interests to do so. Before the rise of rulers and other external managers about 10,000 years ago, cooperative human groups were organised by internal management of this type. Humans generally lived in small bands that were not controlled by a chief or other ruler. The norms and inculcated beliefs that organised these cooperative bands were often entrenched in myths and religious systems.

Internal linear modelling undermines these norms and religious beliefs because it finds no rational basis for them. It is largely unaware of their complex evolutionary function in the formation of cooperative organisation. Instead it sees them as illogical and irrational beliefs that stand in the way of the individual achieving his more immediate physical and emotional goals. The linear modeller is not conscious of the fact that the norms and religious beliefs that are rejected by his limited modelling capacity are adaptations that have essential evolutionary functions.

Once the linear modeller abandons these belief systems, he is left with physical and emotional goals that are largely self-centred. The internal linear modeller is fundamentally ego centric, driven by goals that serve the functioning of the individual rather than the social group. The rise of internal linear modelling amongst humanity has produced the abandonment of religion and the rise of rationalism and individualism that we have seen in the last few hundred years of human history. It is only with the further evolution of the modelling capacity through the development of systemic and evolutionary modelling that the individual will eventually become conscious of the need to reinstate behaviours that produce cooperative organisation. In the next Chapter we will deal in detail with the characteristics of systemic and evolutionary modelling, and look at the superior capacities of self-management that they can underpin.

12

The Self-Evolving Organism

External Systemic Modelling

Linear modellers cannot understand and predict the behaviour of complex systems. Unless a linear modeller can analyse a process into step-by-step chains of cause and effect, he cannot mentally simulate the effects of his actions on the process. This is true whether the process is in his external or internal environment. The limitations of linear modelling mean that there are significant advantages to be had by the development of a capacity for systemic modelling. This ability enables individuals to mentally model their interactions with more complex processes such as social systems and ecosystems. Systemic modellers are able to understand and predict how social systems and ecosystems will unfold over time, and how they might be managed. Individuals with a capacity for systemic modelling are also able to model the effects of their actions over greater scales of space and time.

Because the systemic modeller can understand complex aspects of his external environment, he does not have to simplify them in order to be able to manage them. For example, he can manage and participate in social organisations that are not simplified by rigid codes of behaviour or mechanistic organisational structures. And a systemic modeller does not have to simplify his living environment into a monoculture before he can understand and manage it.

Systemic modelling is made possible by the acquisition of generalised mental schema that represent how various types of complex systems unfold and behave through time. Where the individual has a schema that matches a particular system, he can immediately envisage how the key processes of the system will behave. He immediately sees how the system as a whole unfolds over time, rather than having to follow the interactions of the parts of the system step-by-step.

Unlike a linear modeller, a successful systemic modeller does not analyse a system into its parts and then try to predict the behaviour of the system by seeing how the parts interact together in a step-by-step fashion. Where the mental schema match the system, the systemic modeller will see how the system will behave at a glance, with a flash of insight or intuition.

As systemic modellers improve their ability, they accumulate schema of greater and greater complexity that enable them to model the effects of possible adaptations over wider and wider scales of space and time. They are able to take account of the effects of possible adaptations that linear modellers are completely blind to. But systemic modellers can continue to use linear modelling where it is useful. For example, they still use it where they do not have appropriate schema, and use it as they build up and adapt schema.

Because current science is largely founded on linear modelling, it has great difficulty in accepting and incorporating the findings and insights of systemic modelling. This is the case even where systemic modelling has proven to be indispensable for advancing science. Studies show that few of the great discoveries of science have been produced by linear, logical thinking[1]. A high proportion originated from intuitive leaps made possible by systemic modelling. But before these insights gained scientific acceptance, they had to be translated into simple models based on linear chains of cause and effect. Until this was done, the discoveries were invariably rejected as unscientific.

Importantly, systemic modellers have the potential to manage complex cooperative organisations. They can model mentally how the organisation will respond to their management. They can choose to implement the management that is shown by their mental modelling to advance their interests and those of the organisation. The development of systemic modelling amongst some humans about 10,000 years ago made possible the rise of human communities managed by kings and other rulers[2].

But until the capacity for systemic modelling is turned inwards to further develop the capacity for self-management, systemic modellers tend to pursue the same kinds of values and goals as linear modellers. They are likely to have already developed a capacity for linear self-management. As a result, they probably have analysed and rejected the religious belief systems that were important in organising cooperation within earlier human societies. They will tend to be ego driven and self-centred, and use their enhanced adaptive ability to serve their existing internal physical and emotional goals. They are likely to use the enormous power of systemic modelling to seek narrow goals such as social status, power, feelings of importance, and sexual and other physical pleasures.

Systemic Self-Management

An individual can use systemic modelling to observe and understand his own adaptive processes, and to improve significantly his capacity for self-management. This will produce major adaptive advantages. We saw that linear self-managers are severely restricted in their ability to model the effects of changes to their pre-existing emotional and physical adaptive processes. This is particularly the case for the emotional system. It adapts the individual in complex social situations that cannot be understood by linear modelling. Systemic modelling is not limited in this way. The individual can use systemic modelling to understand the purposes of his existing adaptive systems, and to model the effects of changes to the systems, even where the effects are very complex.

So systemic modelling has the potential to enable the individual to better integrate his mental adaptation with his pre-existing emotional and physical adaptive processes. Using systemic modelling, the individual can begin to manage these adaptive processes to resolve conflicts between them and to ensure their goals are aligned with the goals pursued consciously by the individual. The greater ability of systemic modelling to discover better adaptation will be used to revise the operation of these pre-existing adaptive processes, ensuring that the wider and more complex effects of alternative adaptations are taken into account. Importantly this can include the use of self-management to revise motivations and goals established by the pre-existing internal reward systems. Increasingly, pre-existing motivations and emotional states will be seen as objects of consciousness that can be influenced. The individual will no longer see these as entirely fixed and given, but as increasingly subject to conscious choice.

For example, as an individual's capacity for internal systemic modelling develops, he will learn to recognise a wider range of his feelings and attitudes and understand how they affect his behaviour, and how this behaviour in turn affects others. So that he can improve his interpersonal skills, he will try out different ways of behaving in social situations, with the intention of building knowledge about his emotional responses and their effects. He is also likely to become aware that some of the emotional responses produced in him through his childhood experiences are maladaptive. For example, he may find that his adaptability is restricted because of a fear of change, a need for certainty and predictability in his environment, or an inability to stand up against authority figures even when they are clearly unjust. He is likely to act to revise these inappropriate adaptations. For example, with or without the assistance of others, he may revisit the childhood experiences that produced the maladaptive responses, and use systemic modelling to see what responses would have been more appropriate in the circumstances. He can attempt to revise his psychology so that he would now respond in ways that would be more effective.

A systemic modeller might also attempt to educate his pre-existing emotional responses so that they operate more consistently with the broader understanding made possible by systemic modelling. For example, a linear modeller might tend to find fault and blame in others when they act against his interests, and respond aggressively in anger. In contrast, a systemic modeller may see that the actions of the others were an adaptive response to the circumstances in which they found themselves. The systemic modeller might instead respond by considering how the circumstances that produced the actions might be changed. The systemic modeller might see that blame, anger and aggression might serve no useful functions in these circumstances, and may even be counterproductive.

A systemic modeller might also attempt to ensure that his internal motivation and reward system supports the new adaptive behaviours and strategies that are shown by his modelling to be more effective. He might organise his life and his thinking to ensure he is motivated and emotionally rewarded as he implements these new behaviours and strategies.

Because systemic modellers have a much better understanding of the complex adaptive purposes served by their emotional system, they are also more able to use their emotional states as signals that they should pay more mental attention to particular needs. For example, instead of trying to repress and override feelings of depression, they are more likely to take the feelings as an indication that they need to seriously review their life style. The use of self-management to better integrate the mental and emotional systems means that each system will be used to enhance the adaptive capacity of the other.

As the capacity for internal systemic modelling develops, it will increasingly tend to undermine the individual's self-centeredness. In part this will come about because the individual will begin to see that his particular motivations, goals and values have no absolute value or justification. He will find no valid reason to put them ahead of any other set of goals, and he will be unable to show that a life spent exclusively serving his particular goals and values is inherently better than alternative ways of life. These views will be strengthened as he develops the capacity to model the social processes that have helped to produce his particular set of motivations, goals and values. These models will show that his goals and other adaptive characteristics could have been very different. A different upbringing, different social conditions, a different culture, and he would have different wants and beliefs, and different likes and dislikes. This understanding will begin to undermine the individual's belief that all his energies and adaptive capacities should be solely directed at satisfying his own particular self-centred reward system. It will also help him understand the different perspectives of others, and the causes of those differences. He will be less able to ignore and dismiss alternative perspectives.

Self-centeredness will also be undermined as the individual begins to model the social processes in which he is embedded over wider and wider scales of space and time. He will quickly become aware of his dependence on the effective operation of his social system. He will see that in many respects, he cannot achieve his personal goals unless the social system functions well. The systemic modeller will understand that in many instances, the interests of the social system coincide with his interests, and it is in his interests to promote the effective operation of the social system.

When his models of the social system can span historical scales of space and time, he will increasingly see himself and others as temporary. He will tend to see himself as just one of the enormous number of individuals who make up the social system at any time, and who each follow their particular dreams and goals for the relatively short period of their life. It is only the social system itself that will appear to be able to have any permanence and significance. From this perspective, a life spent solely serving self-centred internal rewards and motivations will appear particularly absurd. Such a life can contribute nothing to anything in the universe that has any chance of continuing in existence, or of having meaning in any broader context. This wider perspective can make it easier for the individual to find value in supporting the effective operation of his social system, or at least the part he interacts with most often. Alternatively, if the individual continues to live a self-centred life, the broader perspective can produce the existential despair that has been common in the 20th century. A wider perspective makes a self-centred existence appear temporary, meaningless, and futile[3].

So systemic modelling will tend to undermine self-centeredness. But it does not conclusively point to a new set of values that the individual should pursue. It eventually undermines individualism, but it does not establish a new set of objectives that can guide the individual. It leaves the individual with ambiguity and uncertainty. A belief in cultural relativism is a typical product of systemic modelling. All values, all goals and all motivations are seen as equally valid. Internal systemic modelling increasingly provides self-managers with the ability to harness their motivations and reward systems to new objectives, but it does not establish what those new objectives should be. This is, of course, the position that systemic modellers find themselves in today.

It is only with the development of evolutionary modelling that humanity can again find individual and collective direction.

External Evolutionary Modelling

The development of a capacity for evolutionary modelling enables the individual to see the effects of his actions over even wider scales of space and time. An evolutionary modeller can model complex systems over evolutionary time scales. He has mental schema that enable him to predict how these systems evolve. The individual can model the effects of alternative actions on the likely future evolutionary success of humanity. He can identify evolutionary trends and future evolutionary events, and use this to see what humans must do to contribute positively to the future evolution of life in the universe.

The evolutionary modeller will see himself and human society as a product of evolutionary processes that have a past, present and future. He will understand that his values, beliefs and other characteristics have been produced by past evolution, and he will know why these take the form they do. The evolutionary modeller will see himself and his society as evolutionary work-in-progress. His mental models will show him that humanity is situated part way along a progressive evolutionary sequence, and he will see the future evolutionary possibilities and challenges that confront us. The evolutionary modeller will see what work he and others must do if humanity is to be successful in future evolution[4].

The evolutionary modeller will be aware that future success for humanity will require the progressive development of cooperative human organisations of larger and larger scale, and of higher and higher evolvability. And he will see that this will require the development of a new psychological capacity in individuals, evolutionary self-management.

But evolutionary modelling will not have a significant effect on the goals and objectives of humanity while it is used only to model the external environment. Until evolutionary modelling is turned inwards to model alternative adaptive processes, it will produce only mental, intellectual knowledge. External evolutionary modelling will not itself change the goals and values pursued by individuals. External modelling enables individuals to find better ways to achieve their adaptive goals, but it does not change those goals. External evolutionary modellers will have much the same goals, motivations and values as internal systemic modellers.

Evolutionary Self-Management

When turned inwards, evolutionary modelling has the potential to build on systemic modelling to provide the individual with a comprehensive understanding of his mental, emotional and physical adaptive mechanisms, the social and evolutionary processes that formed them, and the effects over social and evolutionary time scales of modifying their operation.

But will the individual want to exploit these potentials? Will he pursue evolutionary goals, and use self-management to align his pre-existing adaptive processes with his pursuit of evolutionary goals? Or will the individual continue to serve the internal reward systems that have been established previously by the genetic and social evolutionary processes?

We have seen how the development of internal systemic modelling can weaken the tendency of the individual to put his narrow personal satisfaction ahead of all else. But systemic modelling leaves the individual in no man's land. It is unable to replace the weakened self-centred goals with new values and objectives. Internal evolutionary modelling can do this. It can produce psychological conditions within the individual that will increase the likelihood that the individual will adopt evolutionary objectives.

First and foremost, evolutionary modelling enables the individual to see the absolute absurdity of continuing to pursue his pre-existing goals at the expense of evolutionary objectives. With an evolutionary perspective, the individual will see that his pre-existing goals are flawed and short sighted, the product of inferior and limited evolutionary mechanisms. The individual will know that his pre-existing goals are evolution's inadequate attempt to cause him to behave in ways that will bring evolutionary success. Once he has much more effective ways of consciously pursuing evolutionary success, he is likely to see it as absurd to continue to serve the flawed goals.

Second, evolutionary modelling enables the individual to see that humans do not have any choice about whether or not they will pursue evolutionary goals. Whether they serve the pre-existing goals, or use modelling to consciously pursue the objective of future evolutionary success, they will be serving evolutionary ends. The only choice they have is about how good a method they will use to pursue evolutionary ends. They can pursue evolutionary ends by serving pre-existing goals that were established by inferior evolutionary mechanisms. But their evolutionary modelling will tell them that these goals will not guide them toward evolutionary success in the future. Alternatively they can use a superior evolutionary mechanism to pursue evolutionary goals. They can use evolutionary modelling to identify and implement whatever is necessary to enable humanity to participate in the future evolution of life in the universe.

Third, evolutionary modellers will see that their personal psychological struggle over what objectives they should pursue has a wider evolutionary significance. They will see that their struggle is part of the unfolding of a critical step in the evolution of life on this planet. A similar psychological struggle will be played out on any planet in the universe where organisms become conscious of the evolutionary processes that have formed them and that will determine their future. Evolutionary modellers will see that the way in which humans resolve this struggle will determine the longer-term evolutionary significance of humanity. They will understand that if humanity turns its back on evolutionary objectives and continues to serve the pre-existing goals, we will be evolutionary failures. In an evolutionary sense, humanity would die. We would be irrelevant to the future evolution of life in the universe. For humanity to choose to continue to pursue only pre-existing adaptive goals would be to choose evolutionary suicide and irrelevance. But this would not be a realistic option for a humanity that is capable of evolutionary modelling. Once an individual develops a capacity for internal evolutionary modelling, to reject evolutionary objectives would be as unthinkable as is suicide to an individual who is psychologically healthy.

And finally, evolutionary modellers will see that once they have developed the capacity for evolutionary self-management, the direct pursuit of future evolutionary success will not involve any self-sacrifice. They will be able to find motivation and emotional reward in whatever is necessary to pursue evolutionary objectives. Evolutionary self-managers will manage their pre-existing mental, emotional and physical adaptive processes so as to align their operation with evolutionary objectives. This will mean that pursuit of their managed and modified pre-existing goals will result in the pursuit of evolutionary objectives. Self-management will ensure that all pre-existing mental, emotional and physical adaptive processes will also serve the evolutionary objectives identified by evolutionary modelling.

For all these reasons, individuals with a comprehensive capacity for evolutionary modelling are likely to decide to pursue the development of the psychological skills needed for evolutionary self-management. They will want to improve their evolvability by developing a psychological capacity to manage their mental, emotional, and physical adaptive systems to serve evolutionary objectives. To achieve this, evolutionary modellers will make use of whatever techniques and practices they can find that will assist the development of this psychological capacity. At this stage in the evolution of humanity, we have not accumulated much knowledge about techniques and practices that will help produce this psychological transformation. At present, very few humans develop any capacity to manage consciously their pre-existing adaptive systems. The way we behave is still largely determined by our biological past and our socialisation. These influences produce the likes, dislikes, emotional responses, habits of thought and other predispositions that determine what we do in our lives and how we react in any particular situation. Psychologically, we are immersed in our responses and our habits of thought, and have little independence from them. Very few of us ever develop a comprehensive psychological ability to stand outside these predispositions and reactions, and to consciously choose which of them to retain, and which to modify or discard. We are not yet self-evolving beings.

To date, traditional science has produced very little knowledge about how humans can achieve this psychological transformation. Science presently understands almost nothing about consciousness, let alone about how it can evolve. At this stage in its evolution, science is very effective at helping us understand simple processes in our external environment. But because it relies largely on linear modelling, it has made little progress in understanding complex processes, whether they are external or internal to us. To date, science has almost nothing to say about the issues of greatest importance to most humans—the experience of being, and its meaning.

But some knowledge about the possibility of psychological development has existed amongst humans for a very long time. It has long been known that an individual can develop a psychological capacity that will free him somewhat from the dictates of the external events of his life and from his social and biological past—the individual can acquire a mode of consciousness, which gives him some independence from his emotional and physical states. In most cases this knowledge has been developed and passed on as part of a religious or spiritual system of beliefs and practices. The promotion of psychological and spiritual development has been an explicit part of many eastern religions such as Hinduism, Buddhism and Sufism. And it is at least implicit in many varieties of Christianity. Most of these religions have developed particular practices and activities that are intended to assist psychological and spiritual development. Yoga, prayer and meditation are well known examples.

The spiritual system that comes closest to explicitly articulating practices designed to promote the types of psychological development I have discussed here is the system developed in the first half of the 20th century by George Gurdjieff[5]. He drew on many eastern religions and spiritual practices to synthesise a new system that is specifically directed at producing the psychological transformation of humans into self-evolving beings. But the knowledge about psychological development contained in Gurdjieff's system and in other religious and spiritual traditions is invariably mixed in with myths, metaphors, parables and stories that have little substance. Nevertheless, shorn of these embellishments, there is a remarkable level of agreement within these systems about how we can transform ourselves psychologically to become self-evolving beings.

Most of these systems emphasise the importance of self-knowledge.

To develop psychologically, we must come to know ourselves better. We must be as good at understanding and managing our internal environment as we are our external environment. To achieve this, we must separate psychologically into an observing part and an observed part. Our "I" will be associated with the observing part. The observed part includes our thoughts, self-images, emotional responses, and physical reactions. As the observing "I" develops, it will be able to continually observe our mental, emotional and physical reactions to external events. Our internal processes will become objects of consciousness. The "I" will build a comprehensive knowledge of how we react and how effective these reactions are. The "I" will use this knowledge to model the operation of our mental, emotional, and physical processes so that it can see how they are best managed to achieve the objectives of the "I".

A further critical ingredient emphasised by most systems is that the "I" must develop the capacity to disidentify with these internal mental, emotional and physical processes. When this is achieved, the "I" will no longer live in and be immersed in these processes. It will consider itself separate to them, will not experience itself as them, and will be able to watch them unfold without being part of them. It can stand outside and observe them, just as it stands outside and observes external physical objects and events.

This leads to the development by the "I" of a capacity to manage the mental, emotional and physical systems. Because the "I" does not identify with and is not influenced by emotional states, it can ignore and let go of those that are inconsistent with its objectives. For example, an "I" that has chosen consciously to pursue evolutionary objectives can let go of emotional predispositions that would otherwise cause behaviour that conflicts with its objectives. And it can support and 'go with' emotional responses and motivations that are consistent with its evolutionary objectives. Those that clash with its objectives will be weakened and lose their power, and those that are supported will be strengthened.

An "I" that masters these and other techniques can organise any behaviour that it chooses. It can revise and change any of the behavioural predispositions, habits, likes, dislikes, and preferences that the individual had before the new "I" emerged. The individual will no longer be limited in what he can choose to do by his biological or social past, or by external circumstances. The "I" will operate like a manager of a cooperative group of organisms. Such a manager organises cooperation by supporting members who cooperate and by punishing those who undermine it. In this way, the manager aligns the interests of the individual members of the group with the interests of the group as a whole. In the same way, a fully developed "I" in a self-evolving human manages the pre-existing adaptive processes so that their adaptive goals are aligned with those of the "I". The "I" is also like the visionary CEO of a modern corporation. The CEO manages employees so that their interests and goals are aligned with the longer-term vision that the CEO has for the corporation.

The new "I" will be able to consciously manage the resources and capacities of the individual for the pursuit of evolutionary objectives. It will use its understanding of evolutionary processes and the likely course of future evolution to determine what the individual will do with his life. Evolution will have produced a self-evolving organism. Evolution will no longer have to get the organism to do what is best in evolutionary terms by hard wiring it with internal rewards that are correlated with evolutionary success. The organism will no longer spend its life in the pursuit of emotional rewards that are evolution's indirect way of getting the organism to pursue evolutionary success. By consciously managing its emotional and motivational systems the organism will be able to move at right angles to its biological and social past. The organism will be able to use its own models of the evolutionary consequences of its acts to pursue evolutionary goals directly and consciously.

As evolutionary modelling develops amongst humans, much more effort and resources will be put into the discovery and refinement of practices that will assist the development of the psychological skills and structures needed for evolutionary self-management. The acquisition and use of these skills will produce a mode of consciousness that is increasingly more strategic in its operation. Its primary concerns will be the adaptation of the individual to events and processes that unfold over longer time scales. This evolutionary consciousness will be experienced as a more strategic state of being that is not so buffeted or reactive to immediate events. Like the visionary CEO of a modern corporation, the consciousness will not often be involved in the day-to-day operation of the organisation. The evolutionary consciousness will largely be a spectator or witness in relation to shorter-term events and adaptive processes. The pre-existing adaptive systems, managed as necessary by evolutionary modelling, will continue to adapt the individual in relation to these shorter-term events. And the concerns of the evolutionary consciousness will not be self-centred. They will be more universal, focusing on the effective operation of the social system, the evolutionary success of humanity, and ultimately the successful future evolution of life in the universe. The final allegiance of all beings who attain evolutionary consciousness, wherever they arise in the universe, will be to the successful evolution of life in universe.

* * * * *

In the last three Chapters we have looked at how a capacity for mental modelling can evolve and develop. We saw that mental modelling has the potential to be far superior to previous adaptive and evolutionary mechanisms. Its superiority stems from its ability to use models and simulations to anticipate future events and to adapt the organism to them. The models can be improved in the light of experience throughout the life of the organism. Importantly, when mental modelling is combined with a capacity to transmit adaptive knowledge between individuals, a new evolutionary mechanism is born. The knowledge used to construct and operate models can be accumulated across the generations, producing an evolving culture.

Eventually, organisms will accumulate sufficient knowledge about their environment to model and understand the evolutionary processes that have formed them and that will determine their future. They will see that evolution progresses by producing cooperative organisations of increasing scale and evolvability. Potentially, the organisms could use their modelling and understanding of evolution to guide their own adaptation and evolution. If they could use their modelling in this way, they would no longer need to be controlled by internal emotional and physical rewards—they could work out for themselves what they need to do for evolutionary success, and act accordingly. This would be a major step forward in evolvability, and would enable mental modelling to realise its full potential as an evolutionary mechanism. But initially the organisms will be unable to make this transition. They will be not be able to use their modelling of evolution to guide their adaptation. When the modelling capacity first emerges, it has neither the competence nor the capability to control how the organism adapts. The modelling capacity is grafted onto a fully functioning organism that is already adapted by complex emotional and physical adaptive systems. The modelling capacity does not understand sufficient about these pre-existing adaptive processes to take them over and manage them competently. Its initial attempts to do so are likely to be maladaptive. It will be interfering with complex processes that it knows very little about.

Furthermore, the organisms will not be motivated to use their mental modelling to pursue evolutionary objectives. What they find motivating and emotionally satisfying will be determined by their pre-existing adaptive systems. The pre-existing systems will not reward the pursuit of longer-term evolutionary objectives. The goals of the emotional and physical adaptive systems will have been established by shortsighted evolutionary mechanisms that are blind to longer-term evolutionary needs.

As we have seen, the full potential of mental modelling as an adaptive and evolutionary mechanism will not be realised until it is used to develop a capacity for self-management. Mental modelling cannot fully take over the adaptation of the organism until this capacity is developed. It will be unable to use its understanding of evolution to guide the adaptation of the organism. The first step toward achieving self-management occurs when the modelling capacity is turned inwards. Just as modelling the external environment has enabled organisms to manage and manipulate it, modelling their mental, emotional and physical adaptive systems will enable the organisms to manage these systems. Once the organisms are able to understand how their mental, emotional and physical adaptive systems operate, the evolutionary and other functions they perform, and the consequences of changing them, the organisms will be able to mange them to adapt more effectively. And the organisms will be able to manage their motivational and emotional systems so that they find satisfaction is pursuing evolutionary objectives.

Initially, the impetus for the development of a capacity for self-management will come from the immediate benefits it delivers to the organism. Once mental modelling has accumulated sufficient knowledge, it can adapt the organism more effectively and intelligently than the pre-existing adaptive systems. But the main impetus for the development of a full capacity for self-management will not come until the organisms begin to understand the direction of evolution and their place in it. This enables the organisms to see that their development of a capacity for evolutionary self-management is an important step forward in the evolution of life. They will know that if they fail to take this step, they will be part of a failed evolutionary experiment. The organisms will see their own struggle to develop the capacity as part of the unfolding of a significant evolutionary event on their planet. When they work on themselves to consciously educate, train and manage their pre-existing adaptive systems, they will be aware that they are participating in an important evolutionary advance.

We have seen that the development of a capacity for evolutionary self-management represents a fundamental transformation in the psychology of an organism. Before this transformation takes place, organisms use their capacity for mental modelling to understand and manage their external environment. They use mental modelling to manipulate their environment to achieve the internal adaptive goals established by their biological and social past. But the organisms are unable to use mental modelling to understand and manage themselves. They do not have the knowledge or skills to form complex mental models of their own internal mental, emotional and physical adaptive processes, and of the consequences of changing these processes. They are conscious of using their mental processes to pick the best strategies to achieve their pre-existing adaptive goals. But the organisms do not choose these goals consciously. Their consciousness is largely directed outwards, not inwards. The organisms treat their adaptive goals as fixed and given. Without mental models of their internal processes, the organisms can barely conceive of how they could choose and modify their own motivations and emotional impulses.

When organisms have fully developed the capacity for evolutionary self-management, their biological and social past will no longer limit their adaptive flexibility. They will be able to adapt in whatever ways are needed for future evolutionary success. The organisms will not only consciously choose their behavioural strategies. They will also consciously choose their goals and objectives. They will see their motivations and emotional states as things that are subject to conscious choice. The organisms will no longer unconsciously pursue goals determined by their biological and social past. They will no longer pursue goals established by past evolution, goals that were past evolution's best but flawed attempt to get them to behave in ways that bring evolutionary success. Instead they will use their direct apprehension of what will bring evolutionary success to choose their goals and motivations. Free from their biological and social past, they will be able to use the immense power of consciousness guided by evolutionary modelling to determine how they will adapt and evolve.

Humans have barely begun to undertake this psychological transformation. Our psychology is evolutionary work-in-progress. We are an organism in which the capacity for mental modelling has not yet realised its full potential to take over and improve our evolvability. It is only through conscious psychological effort that we will develop the skills and self-knowledge that will enable us to make this transformation.

This completes our consideration of the evolution of the evolvability of organisms. In the next six Chapters we will trace more systematically the actual sequence of progressive evolution that has unfolded on earth as living processes have formed cooperative organisations of larger scale and greater evolvability. As we do this, we will also examine how the evolvability of societies of organisms (including human and insect societies) has evolved, and how their evolvability is likely to continue to be improved in the future.

PART 4

The Evolution of Life on Earth

Past, Present and Future

13

Evolution of life on Earth

Wherever life emerges in the universe, the potential benefits of cooperation should drive progressive evolution. The broad features of this evolution will be the same everywhere. This is because the potential benefits of cooperation are universal, and the general forms of organisation that enable these benefits to be exploited will be the same wherever life arises.

Life will begin with the emergence of entities that can manage matter to make copies of themselves. The potential benefits of cooperation between these entities will drive the evolution of managed organisations of entities. These organisations will in turn be managed to form cooperative organisations of organisations. This process will continue, progressively forming cooperative organisations of greater and greater scale. Each of the larger-scale organisations formed in this way will be made up of smaller-scale managed organisations, and each of these will be formed of still smaller-scale organisations, and so on. Each organisation will be managed so that its members cooperate in the interests of the organisation. Management will achieve this by ensuring that the members of the organisation capture the effects of their actions on others in the organisation, and on the organisation as a whole. As a result, members of the organisation will tend to treat other members as they treat themselves.

The potential benefits of cooperation will also drive progressive improvements in evolvability and adaptability, wherever life arises.

Organisations that are better at evolving and adapting will be better at exploiting the benefits of cooperation. They will be smarter at discovering new and better forms of cooperation amongst the entities that form them, and at adapting this cooperation as circumstances change.

As we have seen, when life first arises it is likely to evolve through the operation of change-and-test processes that involve competition between organisations. The most competitive organisations will produce more offspring. If their competitiveness is due to features that are passed to their offspring, the proportion of organisations in the population that have these features will increase, and eventually all will have them.

But the first organisations that evolve will not contain change-and-test processes that adapt the organisation during its life, and that pass these discoveries to their offspring. Internal change-and-test processes that can produce evolutionary change across the generations have to be established by a long process of progressive evolution. As this progressive sequence unfolds, internal adaptive processes will be developed that can evaluate the effects of possible adaptations over wider and wider scales of space and time. Increasingly, the internal adaptive processes will be able to assess the future effects of alternative adaptations. Increasingly, they will be able to discover adaptations that are useful because of their longer-term effects.

An important milestone in this sequence is when organisations evolve the ability to communicate with each other about discoveries made by their internal adaptive processes. This enables a culture of adaptive knowledge to be accumulated over the generations. When this milestone is reached, the internal adaptive processes become an evolutionary mechanism that operates during the lives of the organisations.

A further critical milestone is reached when the internal adaptive processes are able to predict future evolutionary events and trends, and can use this ability to discover how to adapt to achieve future evolutionary success. Organisations that develop these abilities will understand the evolutionary processes that made them and that will determine their evolutionary future. As they develop the ability to consciously use their understanding of evolution to guide their own evolution, they will begin to transcend their biological and social past. They will develop the capacity to revise and improve their shortsighted pre-existing adaptive processes to take account of the longer-term evolutionary effects of their actions. They will develop the ability to change their pre-existing adaptations, values and goals so that they can do whatever is necessary to contribute most to the successful evolution of life in the universe. This contribution will include support for the formation of larger-scale cooperative organisations that are managed so that the organisations have the capacity to adapt for the outside/future and to pursue long-term evolutionary objectives.

Wherever life emerges in the universe, its evolution will have these broad features. The scale of cooperative organisation will increase progressively, as will its adaptability and evolvability. But the details will differ. In different locations under different environmental conditions, the nature of the entities that are able to emerge and manage matter to produce copies of themselves is likely to differ substantially. And the molecules that are able to manage cooperative organisations of these first self-replicators may be quite different to RNA and DNA. Even where conditions are identical, the fact that this early evolution will proceed by trial-and-error means that significant differences are still likely. Historical accidents are a common feature of trial-and-error searches in a complex environment.

And any differences that do arise are likely to be magnified. This is because evolution progresses by building on earlier steps. We are managed organisations of the organisations that evolved before us on this planet, and this will be a feature common to all life, wherever it evolves. If any of the earlier steps had been different, due to differences in conditions or historical accidents, all the evolutionary steps that followed may also have been affected, at least in their detail. As a result, organisations elsewhere in the universe that are capable of evolutionary modelling and self-management may look very different to us. But they will be managed organisations of smaller-scale managed organisations of still smaller-scale organisations.

To this point in the book we have concentrated on developing an understanding of the general sequence of progressive evolutionary change that will unfold wherever life arises. We are now in a position to use this general understanding to look at the specific way in which this sequence has unfolded in the evolution of life on earth. In particular, we can now identify each of the organisational milestones in which groups of living entities were managed to form new, larger-scale cooperative organisations. We have already looked in detail at many of these milestones to illustrate how the general evolutionary sequence unfolds. Where we have done so, our consideration of them here will be brief. We will be concerned mainly with putting the milestones into their proper historical order, and looking briefly at how each milestone built on previous steps in the sequence.

Our tracing of this sequence of progressive evolution will eventually bring us to human evolution. We will look in much greater detail at the sequence of evolutionary milestones that have been achieved in the course of human evolution. We will see where humanity fits into the sequence of progressive evolution that has unfolded and is likely to continue to unfold on this planet. This will help us to develop the background needed to see how human organisation must change if we are to continue to participate successfully in future progressive evolution.

The first major organisational milestone in the evolution of life on earth was the emergence of molecular processes that could manage atoms, molecules and other matter to produce copies of themselves. The ability of these molecular processes to manage matter in this way meant that they had a fundamental characteristic of life: the capacity to reproduce. As we have seen, a strong argument can be made out that the first processes to do this were autocatalytic sets of proteins[1]. Collectively, an autocatalytic set of proteins manages the matter available to it to reproduce each member of the set through time. Each member of the set manages (catalyses) chemical processes that lead to the production of other members of the set. Together these managed chemical processes form a proto metabolism.

The next major organisational milestone was achieved through the management of autocatalytic sets and their proto metabolism by RNA molecules. RNA could use the rich molecular resources of a set to boost their own reproduction. Some RNA molecules discovered how to enhance the productivity of their set by catalysing the formation of useful proteins. This enabled the RNA molecules to harvest more from the set for their own use. The most powerful way in which they could boost the effectiveness of their set was to promote cooperation amongst the proteins in the set. The RNA management could ensure that proteins were catalysed to the extent that they contributed to the effectiveness of the set. In this way, proteins captured the effects of their actions on others in the set, and on the set as a whole. The end result of the evolution of this new form of organisation was the first cells. The modern representatives of these simple cells are the bacteria and the other cells known as prokaryotes.

The third key organisational milestone involved the formation of managed organisations of these simple cells. The result was the modern eukaryote cell that contains within it the descendants of the earlier, simpler cells. There is still disagreement amongst biologists about how many different types of cells formed the eukaryote cell. But it is common ground that at least the mitochondria within animal and plant cells, and the chloroplast found only in plant cells, are both the descendants of simple prokaryote cells[2]. These simple cells were initially engulfed by large prokaryote cells as food. But instead of just using them once as food, the larger cells found they could do better by exploiting the benefits of on-going cooperation with the smaller cells. As we have seen in the case of mitochondria, they could do this by managing the smaller cells so that the interests of the smaller cells were aligned with those of the organisation as a whole. Management ensured that the smaller cells could capture the benefits of any contribution they made to the effectiveness of the host cell.

In each these first three organisational milestones, cooperative organisation was produced by external management, rather than by distributed internal management: in the first, matter and a proto metabolism were managed externally by an autocatalytic set of proteins to form a self-reproducing molecular process; in the second, sets were in turn managed externally by RNA molecules to produce simple cells; and in the third, simple cells were in turn managed by the DNA of a larger cell to produce complex eukaryote cells.

Why was it that none of these first three major organisational milestones were produced by distributed internal management? To see why, we need to remember that internal management controls a group of entities by predisposing each of the entities to behave in particular ways. It can do this only if the controls that establish the predispositions are reproduced in each of the entities. Examples of these types of controls are clusters of genes that are reproduced in each member of a group of animals, and clusters of learnt behavioural predispositions such as beliefs and norms that are reproduced in each member of a group of humans. We also need to remember that if a system of internal management is to evolve and adapt, it must include a change-and-test process that tries out different management controls.

This means that if a group of entities are to be managed cooperatively by distributed internal management, the entities must include internal controls that are evolvable and can be reproduced in all members of the group. Internal management could not play a part in producing cooperative organisation until internal controls of this type had evolved.

But distributed internal management was always going to become significant in the evolution of life. This is because progressive evolution was always going to produce internal controls of the type that would make internal management possible. To see why this is so, we need to notice that external managers that organise entities into cooperative organisations will also be internal controls within those new organisations. So the RNA that externally manages an autocatalytic set of proteins to form an early cell is also an internal controller of the early cell considered as a whole. The same applies to the eukaryote cell. The eukaryote cell is formed when the DNA of a larger cell externally manages molecular processes and other simple cells within the larger cell. It is clear that this DNA is also an internal controller of the resulting eukaryote cell, when the eukaryote cell is considered as a whole.

But a further condition had to be met before these internal genetic controls could be reproduced in each member of a group of cells to manage it cooperatively. In order to be able to produce cooperative behaviour between cells, the genetic controls had to be able to control and adapt the way in which the cells interacted[3]. The genetic controls had to be able to organise the cells sufficiently well to be able to produce, control, and adapt useful cooperative interactions between cells.

It was not until the third organisational milestone was reached that this condition was met. It was only with the evolution of the complex eukaryote cell that genetic managers evolved sufficient management ability to produce and adapt the sophisticated interactions that were needed to organise a cooperative division of labour between cells.

This paved the way for the fourth major organisational milestone in which the management of groups of entities produced new, larger-scale cooperative organisations. Highly cooperative organisations of eukaryote cells were made possible by the management of groups of cells by gene-based distributed internal management. A group of cells could be managed by internal genetic predispositions if these were reproduced in each of the cells. This would be the case if the group resulted from the reproduction of a single cell. As we saw in Chapter 7, this genetic internal management could hard wire the cells in the group to act cooperatively in the interests of the group as a whole.

It is worth emphasising here that the multicellular organisms formed in this way were, like the other organisations produced in each of the organisational milestones, organisations of the organisations formed at the previous milestone. These in turn were organisations of the organisations formed at the milestone before that, and so on. So multicellular organisms such as us are made up of eukaryote cells that are the descendants of a community of simple prokaryote cells. These in turn are descendants of autocatalytic sets managed by RNA and DNA, and these in turn are the descendants of the first molecular processes that were able to reproduce themselves by managing simpler atoms and molecules into copies of themselves.

The formation of multicellular organisms again substantially increased the scale over which living processes cooperated. Their larger size enabled multicellular organisms to develop more complex internal arrangements. This made possible the evolution of the organs and systems needed for more sophisticated internal adaptive processes. As we have seen, the genetic managers of multicellular organisms have been able to organise the activities of millions of cells to cooperatively produce complex nervous systems and brains. These have enabled organisms to learn and discover better adaptations during their life, store the results of this learning, and use it to target the search for adaptation in the future. The genetic manager itself was unable to adapt organisms during their life, but it has produced nervous systems and other internal adaptive processes that can.

Genetic distributed internal management also produced the next major organisational milestone. But this time genes organised cooperative groups of multicellular organisms rather than cooperative groups of cells. Ant, bee and termite societies are the most widely recognised examples of this fifth organisational milestone. The genetic managers of multicellular organisms were able to organise societies once the managers had developed the ability to produce and adapt cooperative interactions between organisms. A group of multicellular organisms could be managed by internal genetic predispositions if these were reproduced in each of the members of the group. This could be achieved if the group began with an individual that reproduced to form the other members of the group. A genetic manager produced in this way could predispose the members of the group to act cooperatively. It could also organise them to exclude from the group individuals who were less likely to contain the manager, and to restrict reproduction to those most likely to contain the manager.

Initially, the first societies of organisms organised by genetic managers could not adapt as a coordinated whole during their life. This is because their genetic management could not include a genetic change-and-test mechanism that operated within the society. As we have seen, to maintain full control over an organisation, a genetic manager must suppress competition from alternative managers within the organisation. It therefore cannot try out new mutated managers within the organisation during its life. So, as with multicellular organisms, the genetic managers of insect societies had to establish new adaptive processes within the societies. They had to organise new change-and-test processes that operated during the life of the society.

Insect societies managed by genetic managers have not yet established internal adaptive processes anywhere near as sophisticated as those contained in the more complex multicellular organisms[4]. There is no equivalent to the central nervous system or the brain. This is because insect societies are limited in their ability to explore fully the benefits of cooperation between their members. They have not been able to establish the extensive cooperative division of labour that has enabled multicellular organisms to produce complex organs such as the eye, heart, kidney and brain. Even the most complex insect societies contain only four different types of members[5].

In large part this limitation results from the fact the genetic manager is not reproduced in all the members of a typical insect society. This is because the female insect that founds the society produces the other members by sexual reproduction. The founding female will have been inseminated by one or more males. Offspring that are produced sexually in this way will differ from each other genetically. As a result, the members of the society are likely to have different genetic managers. In contrast, the cells that make up a multicellular organism are each produced by the cloning of a single fertilised egg. Barring fresh mutation, all cells will have the same genetic manager.

So in insect societies, competition between alternative managers within the society will undermine cooperation. It will prevent a manager from capturing all the benefits of the cooperation that the manager could potentially organise. As a result, it will not be profitable for the manager to organise some forms of cooperation even though they would benefit the society as a whole.

As for insect societies, the next two organisational milestones in the evolution of life on earth also involved the formation of organisations of multicellular organisms. But these new organisations were a clear advance over insect societies. Their capacity to explore the potential benefits of cooperation over wider and wider scales of space and time was far superior. This is because the organisations were made up of multicellular organisms that were far more adaptable and evolvable than insects. Organisms that are more adaptable and evolvable are able to form societies that are also more adaptable and evolvable, and are better able to exploit the benefits of cooperation. Smarter organisms can build smarter societies.

The sixth and seventh organisational milestones had to await the emergence of multicellular organisms with highly developed abilities to adapt and evolve during their life. As we shall see in detail in the

Chapters that follow, the sixth milestone required an organism with the ability to learn new behavioural predispositions and to pass them from generation to generation. The seventh could begin to unfold only once an organism with a capacity for systemic modelling had emerged. And the full potential of the seventh milestone will be realised only when there is an organism with a capacity for evolutionary modelling and self-management. Of course, life on this planet began to acquire these capacities only with the emergence of humans and our recent ancestors. The story of the sixth and seventh organisational milestones is the story of the evolution of human societies.

14

Management by Morals

In humans, it is not only our genes that can predispose us to behave in particular ways. Our behaviour can also be controlled by beliefs and emotional reactions that are instilled in us, usually when we are young. Through the process of socialisation, our parents and others inculcate us with ways of behaving that become part of who we are, and influence our actions for the rest of our life. As a consequence, inculcated behaviours, like genes, are able to organise cooperative groups by establishing distributed internal management. In this Chapter we will see how inculcated behaviours produced the cooperative bands and tribes of early humans that represented the sixth major organisational milestone in the evolution of life on earth.

Inculcated behaviours typically include moral codes and social norms that influence the way we deal with others. Norms differ between cultures, but they are likely to include beliefs that it is wrong to cheat and lie, to murder, to fail to return favours from friends, to ignore those in urgent need of help, and to commit adultery. And cultures usually include norms that support and authorise punishment of those who break norms. Commonly, these systems of norms are reinforced by religious beliefs[1].

Like genetic predispositions, inculcated behaviours are resistant to change during the life of the individual. Once a person is inculcated with a system of religious beliefs and norms, they are likely to be followed no matter what the consequences to the individual. The individual will continue to follow norms even though it may mean giving up more immediate biological and social satisfactions. History gives abundant examples of where individuals have been prepared to die for their beliefs. An individual can be as immutably hard wired with inculcated behaviours as with genetic predispositions.

Of course, most behaviours that humans learn during their life are not as fixed and inflexible as inculcated norms and moral codes. Practical behaviours that are used in our day-to-day activities such as preparing food and performing work can be changed and improved continually during a person's life. We are able to adapt these behaviours in whatever way works best. In contrast, inculcated norms are followed even when they do not work well for the individual, and bring nothing but trouble.

But it is the ability of inculcated behaviours to control our actions no matter what the consequences that enabled them to produce a new form of cooperative organisation in humans. This ability enabled inculcated behaviours to establish distributed internal management[2]. Like genes, if inculcated behaviours are reproduced in each member of a group of individuals through time, they can control and organise the group. Inculcated behaviours can hard wire the individual to behave in ways that would not otherwise be in his immediate interests. They can cause individuals to act cooperatively, and to avoid non-cooperative behaviour that would otherwise produce immediate gratification.

Inculcated behaviours also have the potential to organise individuals to punish or expel from the group any individuals who do not apply the behaviours. This ensures that non-cooperators such as free riders and cheats do not undermine cooperation within the group, and take it over. Only individuals that contain the inculcated behaviours capture the benefits of the cooperation organised by the behaviours. As a result, cooperators organised by inculcated behaviours can do better than non-cooperators within a group.

Groups whose internal management is better at promoting cooperation and suppressing cheating will be more competitive than groups with less effective management. Competition between groups will favour those that are managed by clusters of moral codes and norms that are better at organising cooperation. Competition within groups will favour internal managers that are better at ensuring all individuals are inculcated with the clusters of behaviours, and that are better at capturing the benefits of cooperation.

The development in humans of a capacity to inculcate their children with behavioural predispositions enabled the formation of cooperative groups managed by these behaviours. Inculcation was made possible once individuals could learn new behaviours during their life. For its full development, it also depended on the evolution of language.

For most of the last 100,000 years up until about 10,000 years ago, humans lived as foragers in small multi-family cooperative bands of a few tens of people. These bands were typically linked into cooperative tribal societies of a few hundred to a few thousand people. The bands within a tribe met regularly and shared common beliefs and cultural backgrounds. Individuals could move between bands, but only if the bands were within the one tribe. Inculcated moral codes and social norms that were passed from generation to generation controlled the behaviour of the people within bands and within each tribe to produce cooperative organisation. And the codes also organised members of the group to punish any individuals who broke the codes[3]. Unlike the more complex hierarchical human societies that began to emerge about 10,000 years ago, powerful kings or rulers did not govern the earlier bands and tribes. External management played no role in the organisation of cooperation.

A distinctive feature of the codes and norms that organised these tribal societies is that they tended to produce egalitarian behaviour, and sanctioned the punishment of individuals who attempted to compete against others within the group. Sharing of food and tools was required by the codes, and any attempt to monopolise resources, accumulate personal possessions, or to set oneself above other members of the group was generally punished and discouraged[4].

The egalitarian behaviour within these early human groups contrasts markedly with the way in which groups of other apes are commonly organised. Typically, apes compete with one another for status with the group. A disproportionate amount of sex, food and other resources are monopolised by the dominant male. Cooperative behaviours amongst apes are also far less developed than those found in modern tribes of human foragers such as the people of central Australia[5].

The organisation of cooperative groups by inculcated behaviours has significant advantages over the genetic management that underpinned the fourth and fifth organisational milestones. Effective genetic management requires that each member of a group contain the same cluster of genes. But this can be achieved in a group of sexually reproducing individuals only if the members of the group are very closely related, sharing a very recent ancestor. Even then, sexual reproduction will inevitably produce individuals that do not contain the manager. If it is to continue to control the group, the manager must organise the expulsion of these individuals.

This requirement for very close relatedness greatly limits the effectiveness of genetic management. It means that genetic managers cannot organise large groups of sexually reproducing humans: as a group grows in size through the reproduction of its members, the members become less related. They become more distant from a common ancestor and, due to genetic shuffling and mutation, less likely to contain identical managers.

The social insects got around this problem to an extent by reproducing all the individuals in a society from a single fertilised female, the queen. So all the members of the group remain closely related, and are therefore more likely to contain the genetic manager, no matter how large the society grows. By organising the society so that all individuals are nearly genetically identical, the manager is largely able to prevent the emergence of cheats and free riders, and suppress destructive competition. But, as we have seen, there is also a down side to organising a society in this way. It prevents the testing out of new genetic possibilities during the life of the society. A society cannot adapt its genetic management during its life. New possibilities can be tried out only with the formation of new societies, which then compete with each other.

But there is a further down side to the way in which insect societies have suppressed internal competition. The problem is not just that genetic change and adaptation is prevented within the society in the genes that make up the genetic manager. Genetic change and adaptation is also prevented in all other genes. All genetic change is frozen within the society during its life, not just genetic change within management. In order to prevent destructive competition, the manager structures the society to suppress the emergence of alternative managers. But the method it uses to do this also suppresses change (and adaptation) in all other genes, even though they may have nothing to do with management. So, during the life of the colony, insects are unable to adapt the genes that establish their physiology, metabolism, structure, or feeding and other behaviour. In principle, this is a completely unnecessary restriction. Insect societies would be far more adaptable if genetic change-and-test processes could continually adapt these non-management genes in the light of experience during the life of the society, and if only changes in management genes were suppressed.

The fundamental advantage of management by inculcated behaviours is that it can overcome this limitation of genetic management. It can control the aspects of organisms that must be managed to produce cooperative organisation, but without preventing adaptation and change in other aspects of the organisms during the life of the group. Importantly, it can achieve this even in large groups containing individuals that are distantly related. The advantage of clusters of inculcated behaviours is that they can be reproduced in all individuals born into a band or tribe, irrespective of the relatedness of the individuals to other members. Genetic diversity does not have to be suppressed within the group. Nor does behavioural diversity have to be suppressed. Only the particular behaviours that are needed to control the band to produce cooperative behaviour have to be inculcated in all members of the tribe. It is only in relation to these behaviours that destructive competition must be prevented, and change cannot be tolerated. Alternatives can be tried out for all other behaviours, whether they are learnt or gene-based.

As a result, in tribes of early humans managed by inculcated behaviours, sexual reproduction could continue to try out genetic changes without producing destructive competition. And tribal members could continue to test out changes in learnt behaviours that were not controlled by the manager. They could attempt to invent better tools, weapons, cooking methods, and ways of collecting food. It was only behaviours that were part of the manager that had to remain fixed and frozen, and had to be reproduced accurately in all members of the tribe.

As a result, cooperative groups managed by inculcated behaviours could grow larger and live longer than those with genetic management, without having to freeze all other evolutionary adaptation within the group. So human foraging bands and tribes managed by inculcated norms and other beliefs were capable of a much higher level of evolvability than insect societies. Within the tribe, non-management features that were gene-based could be continually adapted. And individuals could continually search for improvements in non-management learnt behaviours, and pass their discoveries onto others, accumulating knowledge across the generations.

The emergence of this new form of distributed internal management enabled a significant increase in the scale and complexity of cooperation amongst early humans. Previously, some forms of cooperation had evolved amongst humans through genetical kin selection and reciprocal altruism[6]. These mechanisms were discussed in Chapter 5. But kin selection, a form of genetic internal management, could produce cooperation only within very small family groups of highly related individuals. As groups grew in size, kin selection was incapable of holding them together, and they would break up into smaller family units.

Reciprocal altruism was limited to producing simpler forms of cooperation, and then only within small groups in which members interacted repeatedly over long periods of time. As we saw in Chapter 5, it is only in these restricted circumstances that it is in an individual's interests to cooperate rather than to cheat by failing to return cooperative favours. In a small group, non-cooperators may lose out because they will be branded as a cheat, and excluded from future beneficial cooperative exchanges. But reciprocal altruism will not work where individuals meet only occasionally. In these circumstances, individuals will not lose much if those that they cheat refuse to cooperate with them again. Individuals will end up in front by taking favours but not returning them, and moving on to do the same to other individuals who have no knowledge of their reputation.

In contrast, inculcated behaviours is able to organise cooperation between individuals who rarely meet. For example, a manager can include behavioural norms that require the individual to cooperate with other individuals who also contain the manager. In these circumstances, even if one of the cooperators gets more of the benefits than others, all the benefits of cooperation will be captured by the manager[7]. But for this to work, the norms must direct cooperation only towards other individuals who contain the manager. The norms must organise individuals so that they identify other individuals who have the same cultural background as themselves, and therefore the same codes and norms, and the same manager. And then the norms must ensure that individuals cooperate only with these others, and not with outsiders, so that only the manager benefits from the cooperation. Ethnocentricity is essential to the evolutionary success of distributed internal managers made up of inculcated behaviours.

A manager could also organise cooperation successfully between individuals who meet rarely if it includes norms that patch up the limitations of the reciprocal altruism mechanism. Such a manager could enable reciprocal altruism to operate more widely and effectively. For example, the manager could include a norm that prohibits cheating in cooperative exchanges. So in a tribe organised by such a manager, an individual could be confident that another person bound by the same codes and norms would return cooperative favours, even though the individual had little knowledge of the past behaviour of the other person in similar situations. But again, the manager would have to ensure that individuals cooperated only with others of the same cultural background.

So internal management based on inculcated behaviours could easily organise families into cooperative bands, and bands into cooperative tribes. It produced the multi-band tribal societies of a few hundred to a few thousand members that appear to have been typical of human foraging groups up to the present day. The members of each tribe shared a common language, religion, norms and moral codes. For most of the time these peoples lived in small foraging bands of a few families. But periodically they would meet in celebrations, mass rituals, inter-band marriages and other collective activities. Individuals could move between band within the tribe, and bands would cooperate with one another, sharing food in the case of emergencies, and joining together to defend against other tribes. In contrast to this cooperation within tribes, societies were generally aggressive toward societies organised by other sets of inculcated beliefs. Members of other tribes were often demonised and seen as sub human, in large part because they followed norms and codes that were different, and therefore unacceptable. To kill another within your tribe made you a murderer, but to kill a person from another tribe made you a hero[8].

Distributed internal management was able to exploit the benefits of cooperation more effectively and over wider scales than kin selection and reciprocal altruism. But it did not replace or overturn the smaller-scale cooperation produced by the earlier mechanisms. Kin selection and reciprocal altruism continued to operate within the larger-scale groups, organising cooperation between closely related individuals and amongst those who lived or worked together for long periods. Many of the behavioural predispositions that evolved to organise this cooperation are still in place today, producing the cooperation within families and inside friendship groups that are still a feature of modern human society[9].

Mechanisms that help ensure that inculcated behaviours retain their control over individuals throughout their life have been very important in the evolution of human groups. A manager can organise a tribe only while it controls the behaviour of the members of the tribe. If its inculcated behaviours lose their control over the members of the tribe, uncontrolled self-interest will reassert itself, and wider-scale cooperation will collapse. The difficulty faced by the manager is that to maintain control it must continue to cause individuals to behave in ways that might otherwise not be in their immediate interests. For example, it must stop individuals cheating in cooperative exchanges when to cheat might be to their immediate benefit. Individuals will be continually tempted to break the moral code and norms. The temptation will be at its greatest in circumstances where breaches could go undetected, and not be punished by the group.

The most successful internal managers would be those that are best at reproducing their behavioural predispositions in all the members of the tribe, and at maintaining their hold over each and every member throughout their life. Managers that failed to do this would be out-competed and fail to survive. The successful managers would be those that organised the process of socialisation to entrench their moral codes and norms in all individuals. Through effective socialisation they could ensure that the individual's internal emotional reward system would punish behaviours that broke norms, and reward actions that obeyed them.

But the hold of the manager over an individual would be greatly strengthened if the individual's mental modelling of the world led him to believe that the codes and norms were in his best interests. Of course, accurate mental modelling of the material world would not lead the individual to this conclusion in many situations. It would often not be in his immediate material interests to follow the codes and norms.

But this difficulty does not apply to mental models that include supernatural beliefs. For example, if an individual could be led to believe that there was an after life, and that he would find happiness in the after life only if he followed the tribe's norms in this life, he would believe it in his interests to follow tribal norms. The same would apply if he thought that individuals who broke norms would eventually be punished in this life by supernatural entities. In this way, supernatural belief systems have the potential to sanctify moral codes and norms, entrench them in a group and ensure they maintain their hold over members throughout their life[10]. They can entrench behaviours that would not otherwise be in the immediate material interests of individuals.

Managers that were able to organise and inculcate religious and mythological belief systems of this type gained a significant evolutionary advantage. They would have been better at maintaining management control over the tribe, and therefore better at exploiting the benefits of cooperation. Tribes organised by such a manager could out-compete other tribes. And the tribes with the most effective religious and mythological belief systems would have been the most competitive. Evolution would have favoured religious and mythological belief systems that were better at entrenching moral codes and norms.

The types of belief systems that were best at inculcating codes and norms would have changed as human knowledge increased. For example, as humans accumulated more and more information about their material environment and how it worked, only those belief systems that were not contradicted by this knowledge could retain credibility and control. The belief systems most likely to survive were those that were largely untestable in the material world, and that could accommodate new discoveries about the world. Belief systems that were more specific and concrete can be expected to have disappeared earlier. For example, many tribes believed that after death, tribal members went to live in the next valley, a far-off island, or some other specific place that the tribe did not visit. Such a belief could not survive direct knowledge of the other place.

* * * * *

In summary, distributed internal management established by inculcated behaviours is able to produce cooperative organisations that are more adaptable and evolvable than those established by internal genetic managers. This is because only those behaviours that must be controlled to establish cooperation are fixed and frozen within the organisation. All others are free to adapt and evolve within the organisation. In contrast, in insect and other societies controlled by genetic managers, all genetic adaptations are fixed and frozen during the life of the society, not just those needed to produce cooperation. Like multicellular organisms, insect societies must evolve completely new internal change-and-test mechanisms if they are to search for better adaptation during their life.

But the evolvability of human tribes and other groups managed by inculcated behaviours is not as effective as it could be. The evolvability of these groups is limited because their management cannot be improved and adapted during the life of the group. As we have seen, the most successful internal managers will be those that organise the group to quickly stamp out potential cheats, free riders and competitors that arise within the groups. Any individual within the group who does not obey all codes and norms could be a free rider who would take the benefits of cooperation without contributing anything to the group. Unless he is brought under control or expelled, such an individual could undermine cooperation, and out-compete the manager.

For these reasons, successful human tribal groups, organised religions and modern cults that are managed by inculcated beliefs are intolerant of deviance from the group's norms. Those who are different are generally hated and demonised. Individuals who do not follow the codes and norms of the groups are punished harshly or expelled from the group. Only managers that organise their group in this way can be successful. But the result is that changes in management that might improve cooperation or adapt the management to changing circumstances cannot be tried out within the group during its life. Like genetic managers, management based on inculcated beliefs cannot include a change-and-test process that operates within the manager during the life of the group. As a result, tribal societies, organised religions, and modern cults are notoriously conservative about their central codes and norms[11].

In principle, a manager could get around this limitation if it were smart enough. It could let changes in management arise within the group, and assess whether each change is harmful or beneficial. The manager could then suppress those that are harmful, and incorporate any changes that are better, improving the group and its management. But like genetic managers, management based on inculcated behaviours does not have the capacity to do this. Within the group, it cannot distinguish between management changes that produce cheats and those that produce better adaptation. To survive it must therefore suppress all changes in management that arise within the group, whether the changes are good bad. It must treat all changes as if they were bad. Moral systems must be paranoid, as well as intolerant[12].

For these reasons, in groups organised by norms and moral codes, new forms of management can emerge only in limited circumstances. For example, changed management may be established when a new group is founded by a small number of deviant individuals who are expelled from or choose to leave an existing group. Or new management may be able to take over an existing group when new behaviours are promoted by a charismatic leader, or in times of great crisis when many individuals in the group are likely to accept that their existing belief systems have failed them[13].

Social organisations would be more evolvable if their managers had the ability to search for improvements in management during the life of the organisation. As we will see in the next Chapter, human external managers who are capable of systemic mental modelling have this ability. Human rulers can continually adapt their management of a society in the light of experience if they are capable of systemic modelling. The emergence of humans with this capacity was essential for the 7th major organisational milestone in the evolution of life on this planet.

15

The Rise of Governed Societies

The human societies that we see about us began to emerge on this planet about 10,000 years ago. It is only since then that governments and other powerful rulers have managed complex human societies, establishing the most cooperative organisations of multicellular organisms to emerge on this planet. Like the complex cooperation found within cells and within multicellular organisms, the cooperation found within these human societies has been made possible by appropriate management, applied in this case by governments or other external managers. Typically these managers have organised cooperation by raising taxes to fund public works, defence forces and other collective activities that benefit the society as a whole. They have also used their power to establish legal systems that punish and deter behaviours which would otherwise undermine economic and other cooperation. In just 10,000 years this management has produced a cooperative division of labour far more extensive than we see in insect colonies or tribal human societies. The division of labour is even greater than we find within our bodies and within cells. The members of a modern nation state are specialised into many thousands of different roles. This immense division of labour is organised to produce knowledge, weapons, technology, food and consumer goods on a scale and complexity unknown to tribal societies.

Since its emergence, the scale of this cooperative human organisation has increased rapidly, from small agricultural communities, to cities, then to nations and empires, and now to global markets[1]. All this has evolved in just 10,000 years, while it took over 500,000,000 years to evolve the division of labour found amongst molecular processes within the eukaryote cell. And it took a further 500,000,000 years or so to evolve the division of labour found between specialised cells and organs within ourselves and other complex multicellular organisms.

This extraordinary explosion in the scale and complexity of cooperative organisation in the last 10,000 years would not have been possible without the emergence of a new form of management. In contrast to the distributed internal management of tribal societies, the new management was external. The members of human societies were no longer controlled only by inculcated behaviours. They were now governed externally by powerful rulers. But it was not the externality of the new management that enabled it to organise human societies that were far better at exploiting the potential benefits of cooperation. It was the ability of the new management to adapt its controls during the life of the society[2]. The ruler could continually try out new controls in the search for improved management. He could adapt his management as circumstances changed. In contrast, the internal management of tribal societies had a very limited capacity to evolve during the life of the tribe. As we have seen, successful internal managers must include behaviours that actively suppress alternative management within the society. If alternatives cannot be tested within the group, internal management cannot adapt or evolve during the life of the group.

The immensely greater evolvability of externally-managed human societies has enabled them to discover much more effective forms of cooperation, including complex divisions of labour. In just 10,000 years societies governed by rulers have ousted tribal societies over the entire planet. This is obviously not because of any innate difference in the capacities of the members of the tribal societies compared with governed societies. It is due to differences in organisation. Governed societies have a much greater capacity to exploit the benefits of cooperative organisation than tribal societies. They are better at discovering ways to ensure that the members of society can capture the effects of their actions on others and on the society as a whole.

Humans were the first organisms on this planet to produce this new form of management because we were the first to evolve a capacity for systemic mental modelling. Systemic modelling is essential if a manager is to be able to adapt his management of a complex organisation competently. This capacity enables the manager to mentally model and therefore to understand how a complex society will unfold through time in response to his management. The use of blind trial-and-error will not enable him to manage effectively. This is because the effects of alternative management actions often will not be felt immediately by the manager. Frequently the effects will unfold only in the future. It will therefore be no simple matter for the manager to determine which effects are the result of any particular management act. Unaided by systemic modelling, the manager will be unable to tell which acts are errors and which are not. As a result, a simple trial-and-error approach will not work. A capacity for linear modelling will be insufficient to overcome this difficulty. The circumstances will be far too complex to be understood by a manager who uses only linear modelling.

Only a manager with a capacity for systemic modelling can plan management changes in a complex organisation, trace the effects of changes when they are implemented, and assess whether a particular change has produced improvements or not. Such a ruler and his associates will be able to construct mental models of how the society unfolds through time, and how it would be affected by alternative management acts. When the effects of an act are different to what was predicted, the ruler can change his assessment of the usefulness of the act, and amend his mental models to improve their predictive ability.

External managers without a capacity for systemic modelling are unable to continually adapt their management of a complex organisation in this way. Examples are the RNA molecules that produced proto cells when they became the external managers of autocatalytic sets. Unlike us, RNA molecules have no internal adaptive processes that enable them to adapt continuously during their life. Instead an RNA molecule adapts by producing offspring, some of which are different and might prove to be better adapted. But, as we have seen, RNA cannot use this process to adapt within a cell during the life of the cell. If it and its offspring reproduce freely in the cell, destructive competition between different RNA molecules will result. So the cell must synchronise the reproduction of the RNA with the reproduction of the cell, preventing competition between RNA offspring within the cell. This ensures that competition can occur only between cells. The only way an RNA molecule can then do better than another is by enhancing the competitiveness of its cell. The interests of the RNA and the cell are aligned.

So the RNA molecules had little ability to adapt their management during the life of the cell. And they had to suppress any change or innovation within the cell that could lead to destructive competition or that could undermine cooperation by producing free riding and cheating. In these respects they were no different to the internal managers of human tribal societies and the genetic internal managers of multicellular organisms and of insect societies.

As we have seen in detail, the inability of these managers to adapt their management during the life of their organisations drove the evolution of new internal adaptive processes. These new internal change-and-test processes searched for better ways to organise cooperation and to adapt cooperation as circumstances changed. A long sequence of progressive improvements in the adaptability of the new internal adaptive processes led eventually to the evolution in complex multicellular organisms of a capacity to use mental models. In humans this culminated in the evolution of an ability to communicate with each other about the knowledge used to produce the mental models, and about the adaptations discovered by the models. This enabled knowledge and adaptations to be accumulated across the generations, and a new evolutionary mechanism was born.

We can now see that this enormous improvement in the internal adaptability and evolvability of multicellular organisms also paved the way for the evolution of societies that were also much more adaptable and evolvable. The critical step was the emergence of an organism with an internal capacity for systemic mental modelling. An organism with this capacity could continually adapt and evolve its management. The emergence of such an organism paved the way for the emergence of organisations whose external management could adapt and evolve continuously. Provided the interests of the management were aligned with those of the organisation, this management would continually adapt its controls in the interests of the organisation as a whole.

On this planet, it was humans that evolved a capacity for systemic modelling, and humans who have produced this new form of organisation. On other planets where life evolves, the organism that develops this capacity might be quite different to us. But the broad evolutionary sequence should be similar. Initially, evolution will produce cooperative organisations whose management is unable to adapt the organisation during its life. The potential benefits of cooperation will drive the evolution of new internal processes that adapt and evolve the organisation. Eventually this will produce organisms with a capacity for systemic modelling. These organisms will be able to form larger-scale organisations whose management can evolve and adapt continually. As we will see, this new form of organisation has the potential to form a unified planetary organisation in which the living processes of the planet are managed to support the pursuit of evolutionary objectives.

The ability of human rulers to manage cooperative social systems has improved considerably over the last 10,000 years. Humans have been able to build better mental models of social systems as we have accumulated knowledge about how the systems unfold through time, and how they react to alternative management acts. As well as producing better management, this has enabled humans to manage cooperative organisations of greater complexity. Government and other rulers have got better at instituting management that ensures members of the society capture the effects of their actions on others and on the society as a whole.

This improvement in systemic modelling has enabled management to be less restrictive and more tolerant of difference. We saw in relation to tribal groups managed by inculcated behaviours that management must be extremely conservative if it is unable to assess whether changes in the behaviour of group members are harmful or not. Such a manager cannot afford to let the members change their behaviours in ways that have any possibility of undermining cooperation. If the manager cannot distinguish changes that have good effects from those that have bad effects, it must suppress all those changes. This seriously restricts the ability of the group to adapt and evolve.

As a result, early human rulers with limited capacities for mental modelling had to manage conservatively. They had to greatly limit the freedom of their subjects to try our new behaviours. Freedom, creativity and innovation had to be restricted to what the ruler and his associates could cope with, given their limited ability to model and understand the consequences of social innovation. In the same way, poor managers in modern human corporations have to suppress the creativity and freedom of bright staff if they are to retain full control over everything that happens under their management. And the built environment humans construct for themselves is simple, predictable and mechanistic so that it is easy to control and manage.

The management task of the early rulers was much easier when the behaviour of the members of the society was already straight-jacketed by effective internal management[3]. As we have seen, distributed internal management could suppress cheating, free riding, theft and destructive competition within a society. An external manager could get away with being far less competent if the society was restrained by a system of mythological and religious beliefs that entrenched and reproduced suitable moral codes and norms. Leading an army to defend the society, organising collective projects such as irrigation schemes, fostering productive economic exchanges and maintaining harmonious and cooperative social relations within the society were all much easier where the behaviour of the members of the society was controlled by appropriate inculcated belief systems[4].

As with the evolution of tribal societies, some systems of myths and religious beliefs were better at underpinning effective external management, and they were favoured in competition between societies. Belief systems were likely to do better if they produced behaviours that made it easier for the ruler to organise effective cooperative activities, legitimated the authority and power of the ruler, and predisposed individuals to behave in ways that benefited society. It was in the interests of rulers and their associates to promote the inculcation and reproduction of belief systems that strengthened their own positions, whether or not they were believers themselves.

As externally-managed human societies increased in scale, producing cooperative organisations of peoples from a number of tribal and ethnic backgrounds, the religions that survived and flourished were those that were more inclusive, and did not preach hatred and aggression against others of different ethnicity. Mythological and religious belief systems that produced acceptance of others from different tribal and ethnic backgrounds produced societies that were able to grow in size, incorporating and managing more and more peoples. Such a belief system could help to produce a cohesive and effective multi cultural society, such as the Roman Empire.

But external management could not rely on the assistance of these internal belief systems forever. As the capacity for mental modelling amongst the members of society improved with the accumulation of knowledge, the consistency of the belief systems with external reality was increasingly questioned. Mythological and religious belief systems were undermined further as individuals developed the ability to turn their modelling capacity inwards to analyse, evaluate and criticise the validity of their own thoughts and beliefs.

Particularly in the last two hundred years, a significant proportion of individuals in more complex human societies has developed a strong capacity to use internal linear modelling to critically evaluate their own beliefs. Increasingly, this has produced a decline in the extent to which individuals are internally hard wired by inculcated beliefs to behave in ways that produce a cooperative and easily-managed society. For more and more individuals, god, tradition and duty are all dead. Their behaviour is now guided largely by internal reward systems that are mostly self-centred, except for the legacy of the kin selection and reciprocal altruism mechanisms. These continue to predispose us toward some cooperation within our families and friendship groups. But this aside, our reward systems tend to promote behaviour that advantages the individual—they largely ignore the effects of our behaviour on others.

This undermining of belief systems that previously helped organise social behaviour has led to the explosive rise in individualism of the last century. Increasingly, it has been left solely to external management in the form of government to manage self-centred individuals in ways that align their interests with those of the society as a whole. In large part, the rise of self-centred individualism has necessitated the massive increase in the size and scope of government that we have seen this century.

As the capacity of rulers for systemic modelling improved during the last 10,000 years, they were better able to cope with diversity and creativity within their societies. Increasingly they could distinguish between changes in behaviour and technology that would benefit society, and those that would harm it. Instead of straight-jacketing society, suppressing all innovation in areas that might be harmful, they could now target their suppression only at harmful changes. Rulers no longer had to rely as much on the assistance of inculcated belief systems to suppress all changes that might be harmful.

Competition between societies tended to favour those that were able to achieve greater adaptability and evolvability in this way. The result has been a potential for greater freedom for members of human society, and societies that are far more creative and innovative. As the capacity of rulers and governments for systemic modelling has improved, human societies have become more evolvable, not only because of the improved ability of management to adapt its controls, but also because it has allowed greater freedom of adaptation amongst the members of the society.

Improved ability to manage diversity and change also enabled governments and other rulers to establish the controls that have allowed large-scale economic markets to arise and flourish. At their heart, economic markets are made up of the same types of exchanges between individuals that underpin reciprocal altruism. An individual gives goods or services to another, and the other reciprocates with goods, services, or money of equivalent value. But the reciprocal altruism mechanism is incapable of establishing the types of exchanges that are essential to modern markets. As we have seen, reciprocal altruism cannot organise exchanges between individuals who are not known to each other and who may not deal with each other again. Rather than cooperate, strangers will pursue their own immediate interests by cheating in exchanges and by stealing. And without cooperative exchange between strangers, modern markets will not emerge.

So the reciprocal altruism mechanism was not responsible for the emergence of large-scale modern markets. Large-scale markets were made possible only by the existence of governments or other rulers. External managers could make market exchanges work by patching up the failings in the reciprocal altruism mechanism. They could do this by establishing a system of controls that prevented cheating and theft. Governments and other rulers typically developed laws and enforcement systems that punished cheating in exchanges, enforced contracts, and prevented theft by establishing enforceable property rights[5].

As we saw in the previous Chapter, systems of inculcated beliefs were also able to patch up the reciprocal altruism mechanism to some extent. But external governance was far more effective. It could implement controls across different cultural groups, and its superior evolvability enabled external governance to discover more effective controls. Importantly, governments were also much better at adapting the controls as individuals found ways around them and as circumstances changed. Without the superior ability of external managers to establish management controls, reciprocal economic exchanges in modern societies would be as limited as those found in tribal societies. Without governments or other rulers, the large-scale complex market systems we see today would not exist[6].

With appropriate management controls, economic markets are extraordinarily good at enabling the benefits of cooperation to be fully explored and exploited in some circumstances. This is because in these circumstances markets enable individuals to fully capture the benefit to others of their cooperative acts. How do markets achieve this? If an individual in an economic market can produce anything that benefits others, he will be able to exchange it for money. If the money exceeds the cost of production, both the producer and the consumer can end up in front from the exchange. The market mechanism creates the opportunity for individuals to benefit by satisfying the needs of others. Individuals can help themselves by helping others. Often they can pursue their own interests best by satisfying the interests of others. Where economic markets are effective, individuals can capture the benefits of cooperating with others. Individuals will treat the other as self.

In a market, individuals have an incentive to search for new ways in which they can satisfy the needs of others. The more innovative and creative they are at producing new goods and services that benefit others, the better off they will be themselves. If members of society have a particular need, a market system will reward the development and production of goods and services that satisfy that need. The market will call the goods and services into existence by making their production profitable.

Competition between producers tends to ensure that the needs of consumers will be satisfied in the cheapest way possible. In principle, any individual in the society can decide to make a particular good or deliver a particular service and try to do it more efficiently than existing producers. A producer who can find a way to manufacture an item more cheaply than his competition will be able to sell it at a lower price while making the same profit. He will increase his total profits as his share of the market expands. In a competitive market, a producer can capture the benefits of satisfying the needs of consumers more efficiently.

The market will organise an efficient division of labour. It does so because the market rewards producers whenever they specialise in ways that produce goods or services more efficiently. A market will also tend to match supply and demand. When demand from consumers for a good is higher than supply, producers can charge more and make higher profits. This will attract more production until supply and demand are in balance[7].

Individuals can participate directly in the market as individuals, or they can be organised cooperatively into firms. Firms are organised cooperatively by a manager in the same way as other cooperative groups. The management of the firm obtains power over employees by giving them wages in return for their submission to its authority. Management uses this authority to organise the individuals to cooperatively pursue the economic interests of the firm.

In summary, modern economic markets have an extraordinary ability to efficiently call into existence an enormous diversity of goods and services that are matched in type and quantity to the needs of the members of the society. And this ability is all a consequence of the capacity of markets to make cooperation pay. Within an economic market, an individual can capture the beneficial effects on others of many of his actions. It is in his interests to do things that benefit others. As a result, markets are able to harness the self-interest and creativity of individuals to produce benefits for all who have purchasing power in the market. Economic markets are one of the most effective ways that evolution has yet discovered to organise cooperation amongst self-interested components.

In this century in particular, markets have proven to be immensely more effective at organising efficient economic activity that the direct actions of government. Centrally-planned command economies such as those in the old Soviet bloc countries have been clearly inferior at satisfying the human needs that markets target. In a command economy, the government and its bureaucracy decides what types of goods and services will be produced and the quantities that will be made, and then organises their production. The inability of command economies to match production to the needs of members of the society was largely responsible for the recent collapse of the old Soviet system of governance.

Why are economic markets so superior to centrally-planned economies? Why cannot a government with all its resources and the ability to employ the brightest people work out the mix of products and services that will best satisfy the populace, and organise production efficiently? The answers to these questions are extremely important. They point to fundamental limitations in the ability of our current forms of government to effectively organise and adapt human society. These limitations impede the capacity of current governments to fully exploit the benefits of cooperation in human organisation, and point to the ways in which governance must evolve if humanity is to fulfil its evolutionary potential.

The most fundamental limitation of centrally-planned governance is that governments and other rulers will always lack the information needed to work out what sort of economic activities are best for the society[8]. A government would need an immense amount of knowledge to decide what range of goods and services would have to be produced to best satisfy the preferences of citizens. Imagine what you would need to know even to work out the quantities and types of shoes that would have to be produced in a complex society to match the diverse and changing needs and preferences of consumers. There is no way that central planners can ever collect the information needed to model a complex society with sufficient detail and accuracy to properly plan economic activity.

Markets do not share this limitation. They leave the final decisions about what goods and services will best match consumer needs in the hands of the only people who know those needs and preferences at any instant: the individual consumers[9]. In a market, producers initially try to work out what goods and services will best satisfy consumer needs and make the largest profits. But the market does not leave these decisions to producers with whatever information they can collect. Instead, through the pricing system the market continually feeds back to producers information about consumer preferences. If prices and profitability are high, the producer increases production. If they are low, the producer decreases production of those goods, and might produce alternative goods. The producers who match consumer preferences most closely and efficiently will generally make the highest profits.

It is important to recognise here that the market mechanism uses a standard change-and-test process to discover the mix of goods and services that will match consumer needs. Like central planners, producers never can know consumer preferences accurately, so they try out their best guesses when they make production decisions. These production decisions are tested by consumers. Production that does not match consumer preferences will be rejected, prices will fall, and production will be changed. This change-and-test process will continually search for the best match between production and consumer needs, and will adjust production to changes in the needs of consumers, the types of goods that can be made, and other circumstances change.

There are three important features of this change-and-test process that are worth noting. Firstly, the process is not internal to humans. It operates through the collective actions of individuals, just as the adaptive processes in our brains operate through the collective actions of cells. It is truly a supra-individual adaptive process. Secondly, the change-and-test process can solve the problem of how to match production to consumer needs even though no human involved in the process might know how to do this. The supra-individual adaptive process can get the right answer even if the producers who happen to be rewarded by the market for best matching the needs of consumers have done so by luck or accident. Like any trial-and-error process, it can get the right answer even if the trial-and-error is completely blind and ignorant.

Finally, the test in the change-and-test process is applied by consumers. It is the purchases made by individuals in a market that ultimately decide the quantity and nature of goods and services that are produced. Through the market mechanism, consumer preferences determine the profitability of each good and service, and thereby control its production.

So markets are much more effective than command economies at exploiting the benefits of cooperation in economic activities. They enable participants to capture the effects of their actions on others, as assessed by those who feel the effects. In contrast, in a command economy the effects of actions are assessed by central planners, who are in no position to judge them accurately. They do not know enough to organise the cooperative economic activities that best satisfy the needs and preferences of consumers. As a result, members of centrally-planned societies will often not capture the actual effects on others of their actions. Things they could do to satisfy the needs of others will not be profitable. The society will fail to fully exploit the potential benefits of cooperation.

The inability of governments to get the detailed information needed to run any complex activity also produces another fundamental limitation of government. A government or other ruler must employ a large bureaucracy if it is to attempt to run a complex system such as an economy. It will need the assistance of many individuals, each of who must develop detailed knowledge of some particular area of the complex system. The government or other ruler cannot accumulate by itself all the knowledge and skills needed to run the economy, and cannot physically do all the things necessary.

The challenge for the government is to get its bureaucracy to do what the government wants them to. It must align their interests with those of the society. If the government fails to do this, the bureaucrats will pursue their own personal interests at the expense of the society. In principle, the government can do this by rewarding bureaucratic work that serves its objectives, and punishing work that does not. But the government is in no position to do this effectively. To do so, it would have to know exactly what actions of the bureaucrats would benefit the system and what would not. The government would have to have a sufficiently detailed knowledge of the circumstances and issues being dealt with by each bureaucrat to independently work out and check what each should be doing. But it does not have this knowledge. If it did, it would not need the bureaucrats in the first place.

It can try to get around this difficulty by developing multi-level bureaucratic hierarchies, with each level attempting to manage the level below to align its interests with the higher levels, and ultimately with the interests of government. But unless a level completely duplicates the work of the level below, it will never have the detailed knowledge necessary to fully and accurately evaluate the work of that level. Employees in bureaucracies always have more detailed work knowledge than their supervisors, unless their supervisors completely duplicate their work. And a level that completely duplicates the work of the level below would not solve this problem. The interests of the duplicating level would not be aligned with those of the government unless it was managed by a level that duplicates its work and knowledge, and so on.

As a result, bureaucracies employed by governments and other rulers are notoriously inefficient at carrying out the tasks set for them[10]. Members of bureaucracies can serve their own interests at the expense of the organisation. Without fear of being held to account by their superiors, they can avoid risk, avoid responsibility, hide behind rules even when the rules are ineffective, work inefficiently, and avoid difficult and demanding work. When they do these things, they can claim that they are doing their job efficiently. Their supervisors will generally not know enough detail about their work to show otherwise. As a result, individuals will not capture the effects of their actions on the objectives of the organisation.

Of course, the same problem applies to any large bureaucracy. The bureaucracies of large multi-national companies are as ridiculously incompetent as those of governments. Corporations spend a large amount of time and resources trying to overcome this problem. For example, they often try to convince their employees to embrace voluntarily values and a vision that will motivate them to act in the interests of the corporation. But most employees find it easier to appear to have internalised company values, rather than to actually do so.

Because governments are unable to align the interests of their bureaucracies with those of the society, the actions of governments and their bureaucracies will often be incompetent. In contrast, effective economic markets align the interests of producers and consumers. Producers capture the effects of their actions on consumers.

It is not only the interests of bureaucracies that are not aligned with those of the society. A further limitation of central planning by rulers is that often the interests of the rulers also do not coincide fully with those of the society they are managing. As we saw in Chapter 5, it will be broadly in the interests of a ruler and his associates (or a government and its supporters) to find better ways to promote cooperation in the society they manage. The greater the productivity of the society, the more the ruler and his associates can take from it. And if the ruler is completely dependent on the society for his future success, he can survive only to the extent he can make it competitive with other societies. The more the society is subject to continual and strong competition from other societies, the more the interests of the ruler and the society will be aligned. The competition will provide rapid feedback to the ruler about his management. If, for example, a ruler significantly increases the share of resources that are taken for consumption by associates of the ruler, he will reduce the competitiveness of the society. If the society is subject to continual and strong external competition, its future will be endangered. A ruler who ignores this feedback does so at his peril.

When human societies were small, numerous and continually competing with one another, competition provided rapid feedback to their rulers. But as human societies have increased in scale, competition has become more intermittent and often less intense. Empires have been able to escape strong competition for long periods of time. Whenever external competitive pressures are weaker, the alignment of interests between the ruler and the society is lessened, and the pursuit by the ruler of his interests may be inconsistent with the interests of the society. The ruler will no longer capture all the effects of his actions on the society. He can get away with using more of the resources of the society for the personal benefit of himself and his favoured associates. He can damage the effectiveness of the society by requiring citizens to work and behave in ways that benefit him rather than the citizens and the society. If external competition is weak, a government can get away with diverting more of the resources of the society to its close supporters and associates.

Weaker external competitive pressures also mean that a ruler who does not use his power competently can cause a lot more damage to the society before competition makes his continued rule impossible. The societies that were incompetently managed by Hitler and Stalin were eventually destroyed by competition. But their mismanagement caused enormous misery and suffering before competition ended their rule.

It is now widely accepted that markets are far superior to centrally-planned economies at producing efficient economic outcomes in many circumstances. We can thank the obvious failures of the communist command economies for proof of this. But before this evidence emerged, many economists and other thinkers found it very hard to recognise the superiority of markets[11]. They saw that if they accepted that government planning was inferior, they must also accept that the power of the human intellect is fundamentally limited. This is because economic planning involves the use of the human intellect in a standard and systematic way. Economic planners proceed by identifying the outputs wanted from an economic system, collecting the information and data needed to understand the system, and using this to work out the best way to intervene in the system to produce the desired outcomes. In general, this is the way we use our intellects to attack any complex problem. It is how humans have gone about attempting to conquer nature. Most of us think it is the best method for solving any complex problem, and it is what we think we do better than other organisms. But what many found very hard to accept is that this standard use of the human intellect is far less effective than a market system at solving the economic problem of matching production to consumer needs. The human mind is clearly inferior to market processes even though markets are not conscious and have no brain.

From where does the intelligence of the market come? As we saw earlier, the superior ability of the market to solve economic problems comes from its use of supra-individual change-and-test processes. These processes use trial-and-error in addition to the knowledge of participants to solve complex adaptive problems. Markets can solve economic problems even though they do not contain the knowledge to build a model that could be used to work out a solution. Change-and-test processes are particularly superior to mental models where there is insufficient knowledge to build an accurate model. Because of this, the best that humans can do to solve complex social problems where there is uncertainty and incomplete knowledge is to set up supra-individual change-and-test adaptive processes such as a market. The use of the intellect alone will fail, as central planners have demonstrated. In these circumstances, governments and other rulers have found that they can promote cooperation better by establishing markets than by direct intervention. Markets are better than calculated and planned interventions at ensuring that citizens can capture all the effects on others of their actions.

* * * * *

To summarise the main themes of this Chapter, external management by governments or other rulers has proven far more effective at organising human societies than internal management by inculcated beliefs. The evolvability of societies that are externally managed by rulers capable of systemic modelling is vastly superior. Human societies managed by governments or other rulers have been able to discover and establish far more complex and effective cooperation between their members, and to adapt their management as circumstances change. In the 10,000 years since external managers first emerged, their management ability has improved significantly. But governments and other rulers are still limited in their ability to directly organise cooperation. They lack the information to plan and implement the detailed interventions that would be necessary to ensure that members of the society always capture the effects on others of their actions. Governments lack the ability of economic markets to achieve this in many circumstances. The best way in which a government or other ruler can manage a society in these situations is to establish a system of controls that will enable economic market to emerge.

But markets have their own serious limitations. We will see in detail in the next Chapter that markets are not able to exploit the benefits of cooperation fully in all circumstances. In fact, in some situations they undermine and prevent effective cooperation. We will see that if a government or other ruler restricts its management to establishing the controls that enable markets to operate, the full evolutionary potential of the society and of most of its members will not be realised.

16

Limitations of Markets

In a range of circumstances, markets are immensely superior to government planning at organising economic cooperation. Markets effortlessly organise a highly efficient division of labour between members of the society. But markets will produce cooperation only to the extent they can ensure that individuals (including firms) capture the effects on others of their actions. All effects on others, beneficial and harmful, must be captured. If individuals are not rewarded for all the beneficial effects of a particular action, the action will not be as profitable as other actions that are less beneficial, and the society will miss out on useful cooperation. And if individuals do not bear the full cost of their harmful actions, they will not be deterred from doing things that harm others.

In what circumstances will a market fail in this way? A free market can work effectively only where the full benefits of goods and services go solely to the purchaser, and cannot be enjoyed free by anyone else who does not pay for them. Unless this condition is met, the producer cannot obtain payment for all the benefits he creates for others. In the case of consumer goods such as refrigerators, washing machines and televisions, this condition is clearly met. The purchaser has the sole right to use the goods, cannot gain the full benefits of a good without purchasing it, and is prepared to pay a price that reflects those benefits.

But not all goods and services are like this. Some goods and services have collective effects that cannot be targeted only at those who pay for them. They have effects on many others, and it is impossible to prevent people from enjoying these effects even though they have not paid for them. It will be in the interests of individuals to free ride on these goods and services by taking them without paying. As a result, the producer will not be able to capture payment for all the beneficial effects on others of his goods or services. Goods and services that would have been profitable if all their effects were captured will not be profitable, and will not be produced in a society that relies only on markets[1].

A classical example of a service with collective effects is the defence of a country. If an entrepreneur organises an army to defend his country against an invader, he will not be able to capture the full benefits of his actions. He will defend all members of the society, whether or not they are willing to pay him for it. There is no way he can parcel up his service so that only those who pay for it are defended. So it will be in the interests of individuals to decline to pay, and to free ride on the service paid for by others. Because entrepreneurs cannot capture all the benefits of providing defence services, a market alone is unable to produce an adequate defence system. The free rider problem can be overcome only by a government or other ruler who has the power to compulsorily raise enough taxes to properly fund defence. Only a manager can ensure that all who benefit from defence pay for it through compulsory taxation.

Another very important example of services that will not be supplied adequately by a free market alone are education and training. The reason for this is that education and training does not benefit only those individuals who consume it. Educated individuals produce collective benefits. This is true both for general education as well as for work-orientated education and training. For example, a democratic society will operate more effectively if its members are well educated about how the social and economic systems function. And work-orientated training benefits not only the individual who is able to earn higher wages with his improved skills, but also all those who work with the individual and who are more productive because they are working with someone with higher skills. So suppliers of education and training will not be able to capture the full benefits that their service provides to society. Free riding will prevent them from capturing benefits that spill over to others. As for defence, this will undermine the provision of an adequate level of education and training unless government intervenes[2].

This failure of free markets to organise an appropriate level of education for members of the society is made significantly worse by a number of additional factors. In a free market, many individuals will not earn sufficient money to buy even the education that is in their own best interests, will be too young when education is of greatest benefit to them to make an informed decision about its value, and will not have sufficient education or information to make sensible educational choices. As a result, a society that leaves the provision of general or work-orientated education to the market will be significantly less productive as a whole. It will be less competitive economically than other societies that educate their members better.

Markets alone will also fail to establish programs and services that can improve the quality of life of all who live in a community or society. Policing, the provision of safe and well-planned public areas such as parks and streets, and programs that reduce crime by rehabilitating drug addicts or by providing satisfying activities for teenagers all have beneficial collective effects that cannot be targeted only at those who are willing to pay for them. No matter how beneficial these sorts of programs may be to a community, a free market will not provide them. A neighbourhood, a city, or a region can disintegrate socially, and the market system will not do anything about it.

Market systems alone will also fail to produce the best outcomes when individuals do not capture all the harmful effects of their actions. Again, this will occur most commonly when the production and use of goods and services have collective effects that go beyond the producer and the consumer. For example, the production of a good might have harmful effects on the environment, disadvantaging many others. But there is nothing in a free market system that requires the producer to compensate those disadvantaged by the pollution he causes. Neither he nor the consumer will pay the price for these harmful effects on others. As a result, the market does not give them any incentive to limit these harmful effects[3].

In fact, if the producer can manufacture an item more cheaply by a process that degrades the environment, he will have to use this process if he is to stay in business. A producer who is bound by a belief system that prevents him from harming others and who therefore refuses to pollute, will go out of business. He will have to use more expensive non-polluting processes and charge higher prices, and will be out-competed in the market place.

It is critically important to note this destructive feature of the way markets operate. It is not just that economic markets fail to produce optimal outcomes in circumstances where they fail to ensure that individuals capture the benefits or harms of their actions. They also actually prevent individuals from behaving in ways that would correct these market failures. Individuals or firms who try to do so will be out-competed and go out of business. The competition that ensures that consumers get goods and services at the cheapest price is a two-edged sword. It destructively ensures that individuals and firms cannot take into account the factors that the market does not, even where it would advantage the community or society greatly if they did so. A firm cannot stay in business in a perfectly competitive market if it refuses to pollute, or if, as a contribution to society, it educates its employees beyond what is needed solely for the operation of the firm, or if it implements programs that produce enormous collective benefits to the community as a whole. Effective free markets make morality impossible. And the only individuals or firms who are able to accumulate great wealth in a competitive market are those who pursue only their own financial interests, and are not prevented from doing so by any conflicting moral code or religious beliefs.

A further major limitation of economic markets is that they satisfy the needs only of those who have money. Markets have an extraordinary ability to efficiently meet the needs and desires of those who have purchasing power. But they will provide absolutely nothing to those who have none. If you do not have sufficient money to buy the food you need to live, the market will do nothing to help you as you starve to death. If you have a curable but potentially fatal illness, and do not have the money to pay for a cure, the market will let you die. Markets do nothing to stop millions dying in these circumstances around the world every year, even though there are enough resources to prevent their deaths. The market's invisible hand is drenched in blood.

Markets give nothing to those without purchasing power because individuals do not capture any beneficial effects they might have on individuals who cannot pay. There is no profit in providing goods and services to those without purchasing power. Participants capture only the beneficial effects they have on those with money. The market will provide great rewards for an individual who builds a factory and employs hundreds of workers to produce tasty new flavours of ice cream for the children of the wealthy. But the market will not provide a loaf of bread to a starving child whose parents are penniless.

This limitation would not be so serious if markets operated to ensure that all members of a society have sufficient purchasing power. Individuals and the society could still reach their potential if the market ensured all individuals had sufficient money for their material needs and for the education they need to make the best possible contribution to their community and society. But markets do not do this. Instead they tend to produce a distribution of income that concentrates most of the wealth of the society in the hands of a small minority. For example, in the United States of America, the top 1 per cent of the population has about 40 per cent of the total net worth[4]. And this is after the US government has redistributed market income through taxes, minimum wages, welfare programs and other transfers.

In large part markets produce this massively unequal distribution because the income of most workers and producers is driven down by competition in the market. Great wealth can usually be accumulated only by those who find a business opportunity whose profitability is not forced down by competition. Or by those who find employment that is sheltered from full competition.

Competition for work drives down the income for most jobs. When there is an oversupply of potential employees for a particular type of work, and the alternative is unemployment, competition will drive wages down to subsistence levels, and working hours will expand. Those who might miss out on work will be willing to accept less to obtain a job. In a perfectly competitive market, the level of wages for a particular type of labour will tend to equate to the cost of reproducing the labour. For workers without special skills, this means wages that are just enough to feed and clothe them so that they can continue working. This was the outcome during the industrial revolution before governments legislated minimum wages and the maximum hours that could be worked for those wages[5]. Even the salary levels of jobs that require extensive university qualifications will be driven down by competition if there is free access to obtaining the qualifications. When there is an undersupply, higher salaries will attract more people to get the qualifications. But once this satisfies demand, competition will tend to force salaries down to the minimum level needed to reproduce the employees and their qualifications.

Competition also tends to produce work that is meaningless and unfulfilling for workers. Competition amongst businesses will force employers to organise work and design jobs to maximize efficiency. They will not be able to structure work in ways that produce satisfying and meaningful work for employees where to do so would conflict with maximum efficiency. Competition between potential employees will mean that they have to accept what is offered. The result is that large numbers of employees in market economies spend their working life doing tasks in which they find no meaning, that are achingly boring, and that prevent them from developing their potential as human beings. Many workers in modern market economies have less freedom in their life to pursue their own interests and personal development than slaves in earlier times. Large parts of their life are taken up doing things that they would never freely choose to do[6]. This is an inevitable outcome of ungoverned (unmanaged) markets.

Competition amongst the producers of goods and services drives down their profit margins. No one makes much money out of businesses that are subject to strong competition. In areas of business where anyone can get the knowledge and capital to operate, competition in a free market system will force down profits to subsistence levels. This is commonly the case where there is no regulation of taxi services, corner shops, street vendors and other small businesses. It is only in circumstances where vigorous and open competition does not exist to drive down profitability that great wealth can be accumulated and that the massively unequal distribution of wealth found in modern market economies emerges. It is only when there are business opportunities that most people are unable to exploit that very high incomes are produced.

For example, professionals of various types have been able to obtain high incomes by building themselves shelters from free and open competition. If a licence is required to practice the profession, and barriers are put in the way of gaining the licence, such as by limiting the number of training opportunities, competition can be avoided and high incomes assured.

The wealthy have much greater access to economic opportunities that are sheltered from competition. As a result, it is much easier for the wealthy to perpetuate their wealth or to increase it. Wealth opens up many income-earning opportunities that are not open to those without wealth. For example, it enables a person to get involved in much larger-scale business opportunities because they have much easier access to finance[7]. And the wealthy are able to purchase exclusive educational opportunities for their children that shelter them from strong competition. The great majority of people are in no position to compete with the wealthy for these opportunities. As we will consider in detail later in this Chapter, the wealthy can also use the power that goes with their wealth to get governments and other rulers to manage the society in ways that favour them, including by further sheltering their businesses from competition.

In these cases, it is clear that great wealth can be produced and maintained only because the market fails to work effectively. If the market could organise competition successfully in these circumstances, incomes and prices would fall, and the needs of consumers would be satisfied more efficiently.

Great wealth can also be acquired by those who are able to produce better goods and services, or can produce the same goods and services at much lower prices. If other producers are not able to duplicate quickly and easily what they are doing, there will be no effective competition in the marketplace to drive down profitability.

For example, a manufacturer may be the first to invent an innovative labour-saving kitchen appliance. Before others are able to develop the capacity to make the appliance, the manufacturer is able to earn very high profits. He escapes the competition that would otherwise drive prices and profitability down. Those who do accumulate great wealth in this way, and those who attempt to justify all aspects of the market, often claim that it is due to special talents and hard work that individuals find these golden opportunities. But this ignores the numbers of equally-talented and hard-working individuals who never stumble on such an opportunity. And it ignores the failure of governments to pick and develop economic winners, despite all their resources and access to expertise. Talent and hard work are not enough. There is a significant component of luck in whether or not an individual finds himself in the right place at the right time with the right knowledge and with the right connections to exploit an opportunity that will make him rich. In fact, it is largely because there is a great amount of unpredictability, lack of relevant information, and need for trial-and-error that those who do find these opportunities can escape competition for a sufficient period to accumulate great wealth.

Of course, a major strength of the market is its ability to reward those who discover new and better ways to satisfy the needs and tastes of consumers. But the pure market does not contain an effective mechanism to ensure that the level of the reward is cost effective. Often the rewards are far in excess of what would have been needed to encourage the individual to make and develop the discovery. Bill Gates would no doubt have settled for far less than 60 billion dollars to build up his Microsoft computer software empire. Not only could the market be far more efficient if it included processes that ensured rewards were not excessive, but it would also produce a more equal distribution of wealth. Again, we find that the accumulation of very high incomes is a product of market failure. In this case it results from the inability of the market to organise sufficient competition to ensure that innovations are produced efficiently at the lowest cost to consumers[8].

The extraordinarily unequal distribution of wealth produced by markets is a result of fundamental flaws and failures in the effectiveness of market processes. These are serious deficiencies. The concentration of wealth produced by these flaws has major disadvantages for the competitiveness and effectiveness of the society as a whole. First, it does not make full use of the potential of many members of the society.

Many individuals will never have the opportunity or the purchasing power to develop their own or their children's full potential to contribute to the effective functioning of society.

Second, this unequal distribution of wealth organises most of the productive capacity of the society to serve the desires and tastes of the small minority who have most of the wealth. What a market does very efficiently is satisfy the needs of those who have purchasing power. And if purchasing power is mostly concentrated in the hands of a few, it is largely the wants and whims of these few that the entire economic system will serve. The market will organise the majority to spend their working lives serving the interests of a minority. Technological advances will be organised by the market to satisfy the interests of the extremely wealthy. This is not a noble outcome for a society. And as we shall see in detail, it will not produce a society that will achieve longer-term evolutionary success.

The tendency of market systems to concentrate great wealth and economic power in a small number of individuals and corporations results in a further major limitation in the effectiveness of markets. Enough wealth can give individuals and corporations the power to influence the government or other ruler who manages the society and the market. As we shall see in detail later in this Chapter, even in democratic societies the wealthy are able to play a major role in determining who gets elected to govern and how long they stay in office. This power gives the wealthy the potential to bias the management of the market so that it operates in their favour. Markets include the seeds of their own destruction. They tend to produce individuals and corporations with the power to undermine the effective operation of the market.

The power to influence the government or other ruler opens up immense new opportunities for wealthy individuals and corporations to pursue their own interests. To be able to manage a society effectively, governments must be very powerful. They must be able to raise taxes, use those taxes to promote particular activities in the society, make laws, and punish anyone who does not obey the laws. As we have seen, this power can be used to organise a cooperative and productive society. But it can also be used to enhance the interests of particular individuals and corporations within the society. Individuals and corporations that accumulate sufficient economic power are able to ensure that the power of government is used to benefit themselves. They are able to use the power of government to ensure that the benefits they receive from society are far in excess of what they contribute. They will capture benefits far in excess of the effects on others of their actions. As a result, the effectiveness and efficiency of the society will be undermined.

There are many ways in which economically-powerful individuals and corporations can use the power of government for their own benefit. They can get the government or other ruler to grant and enforce a limited number of exclusive rights to exploit particular business opportunities or to provide particular services. For example, governments often place restrictions on who can use land in particular ways, set up television and radio broadcasting stations, enter particular professions, catch fish commercially in certain areas, run lotteries and casinos, and import particular types of goods. Because the people given these opportunities are protected from full competition, they can make above-normal profits.

Economically-powerful individuals and corporations can get governments to refrain from breaking up monopolies and from promoting competition in areas where the market is not operating efficiently. For example, in areas of the economy where large-scale businesses have entrenched themselves, it can be very difficult for new, smaller businesses to enter the market. Because competition is limited, high profits can be made. The market would deliver products to consumers at much lower prices if competition were stronger. Governments can do a lot to open these areas to greater competition. But economically-powerful individuals and corporations who can influence governments can protect their high profits by ensuring that the government does not do this. Most modern governments have laws and processes that they claim are directed at preventing monopolies and anti-competitive practices, but these are largely ineffective[9].

More directly, economically-powerful individuals, corporations and interest groups can influence governments to give them special tax breaks, more lenient controls on pollution, exemption from laws that set minimum wages and safety requirements, subsidies, and protection from competition by imports. And they can get the government to ensure that the tax system leaves them with many opportunities to avoid tax. Modern governments typically claim that their tax system collects a higher proportion of the incomes of the wealthy. A tax system that does this is essential if governments are to counter the tendencies of markets to concentrate most of the wealth in the hands of a small minority. But the wealthy can use their influence over governments to ensure that the tax laws that are claimed to do this are, in practice, ineffective and unenforceable. Typically in modern countries, the wealthy with their accountants and lawyers can find and exploit loop holes in the tax laws faster than governments close them. Increasingly, modern governments have used this as a justification for moving towards flat rate taxation systems such as consumption taxes. But there is no technical reason why governments cannot reform their legal system in ways that ensure that the wealthy pay the tax it is claimed they should. However, it is not in the interests of governments to do this.

The usefulness to the wealthy of a legal system that can be manipulated by costly lawyers and accountants points to a more general way in which the economically powerful can bias market systems in their favour. If the legal system that enforces the framework of laws that govern markets is very expensive to use, and if persons are more likely to get a favourable result the more money they spend, the legal system will be less effective at regulating and controlling the behaviour of the wealthy. If the economically powerful are able to escape the framework of laws, they will no longer be restricted to pursuing their self-interest in ways that benefit the society. And they will have a competitive advantage over the majority who cannot escape the laws.

A legal system also will be biased toward the economically powerful if the law is extremely complex, there are many opportunities for appeal, the outcome of each step in the legal process is uncertain and unpredictable, and the best legal advice and representation is very costly. Each of these characteristics make it easier for those with the most financial resources to achieve a favourable result from the legal system, and harder for the majority of citizens to do so. With such a system, the economically powerful do not have to bias the legal process in their favour by risky bribery of the judiciary. Instead, the complexity and expense of the system will ensure that only they can afford to do the things that are likely to deliver a favourable result.

It is in the interests of the wealthy for governments to be small, and they can use their influence over governments to achieve this. The less governments spend on welfare and general education, the less tax they will have to raise off the rich. And the wealthy do not need to rely on the services provided by government. It is in their interests to have governments provide the minimum level of support consistent with ensuring there is no widespread dissatisfaction in the society that could threaten the status quo. The level of dissatisfaction they could get away with was much lower when communism competed with capitalism. The possibility of war meant that it was in the interests of the wealthy to ensure that their societies were sufficiently cohesive for sufficient citizens to be willing to fight and die for it. The economically powerful were prepared to support substantial welfare and social security programs in capitalist countries. Now that communism is no longer seen by many as a realistic alternative, these programs are being dismantled throughout the world, and the wealthy are pushing for small government[10].

The benefits to the economically powerful of co-opting the power of governments and other rulers in these ways are enormous. Throughout history, alliances with rulers have always been the royal road to riches for those who would be wealthy. It has been far easier to accumulate and preserve great wealth in this way than by continually out-competing others in a fair and vigorous market.

Somewhat paradoxically, the same self-interest that makes markets so efficient also motivates the economically powerful to use the power of governments to undermine markets. As we have seen, participants in a market do not have to care about the interests of others to want to strive to produce goods and services that satisfy the needs of others. Because the market enables them to capture the benefits of their effects on others, self-interested participants will be driven to satisfy others, even though they care only about their own interests. The market can organise complex cooperation even though participants have no interest in being cooperative or in making the market system work. Participants just pursue their own narrow interests wherever they lie, wherever that takes them.

But this strength of the market system is also the source of its greatest weakness. Individuals and corporations that are ruthlessly self-interested will flourish in a market system and will benefit others as they benefit themselves. But when they accumulate sufficient economic power to manipulate governments, they will do so. The fact that this will prevent the market from efficiently exploiting the benefits of cooperation will be immaterial to them. They will not give a dam. Cooperation and the interests of others are not their concern in their business activities. They will not be interested at all in whether or not they undermine the market. Their only consideration in planning where to direct their energies is how to gain the most benefits for themselves. If this is achieved best by manipulating market rules, that is what they will do. Whether they put their energies and talents into following the rules that regulate the market, or by circumventing or biasing the rules, depends only on which alternative produces the greatest returns to themselves.

In fact, individuals or corporation that do not behave in this way will be out-competed in the market, unless they can escape competitive pressures. As we have seen, producers will be competitively disadvantaged if they use their resources to satisfy the interests of those who are ignored by the market, such as citizens who have little purchasing power. For similar reasons, they will also be out-competed if they fail to take advantage of opportunities to use government to bias the market in their favour.

But it is not in the interests of the majority of citizens in a society to allow the economically powerful to manipulate governments and other rulers in this way. The greater the share of the wealth of a society that is taken by the economically powerful, the less there is for the majority. And the manipulation of governments by the wealthy to advance narrow interests undermines the broader interests that governments have in promoting the productivity of the society as a whole.

Individuals and groups who have the capacity to use mental models to understand how their society is manipulated by the wealthy are able to see that it is in their interests to put a stop to it. But to end the manipulation, they have to acquire sufficient power to overcome the power of the ruler and of those who support him. Throughout history, citizens have usually been able to achieve this only when large numbers of them have been willing to band together to overthrow the ruler by force. But individuals have not generally been prepared to risk their lives in this way unless they had little to lose and unless they believed that the revolution could install new rulers and perhaps a new way of organising the society that would significantly advance their interests.

These conditions were not often met. It is in the interests of rulers and those who benefit from their rule to manage the society in a way that makes successful revolt unlikely. A ruler could ensure that poverty would not be experienced by sufficient numbers of the society to threaten his rule. And the ruler could promote ideologies and beliefs that suggested that his rule was the best or even the only practicable way of organising an acceptable society, despite any poverty, misery, and inequality it produced[11]. To the extent that these beliefs were promoted successfully, the mental models of members of society would not enable them to foresee any better way of organising society. And the more widely these beliefs were accepted, the greater the inequalities in wealth that could be suffered without risking revolution. The ruler and his associates could take a greater share of the wealth of the society without endangering their continued power.

But the task of promoting these beliefs became more difficult as members of many human societies developed an improved ability to model and understand more complex processes and systems. As humans accumulated knowledge and shared it through books and education, more individuals were able to mentally model how their society functioned, and could see how it might be organised differently. Increasingly, the members of many societies were able to see that there was no basis for the doctrines that were used by their rulers to justify their absolute power. Their rule was not divinely sanctioned, and they were not infallible. And more and more could see reasonable ways in which their rulers could be controlled to ensure they ruled in the interests of the majority. Revolts and threats of revolts progressively resulted in more effective restrictions on those who managed and governed many human societies. Increasingly, these restrictions realigned the interests of the ruler with those of the ruled, and made it more difficult for narrower interests to manipulate the power of the ruler for their own ends.

The most significant organisational innovation developed to date by humans to align the interests of government with the interests of society is democracy. Through typical democratic processes, the majority of the members of a society are able to install a government that they think will serve their interests best. And they can throw out the government in three or four years if they think there is a better alternative. The need for a violent revolution to overthrow a government that does not act in the interests of the majority is replaced by regular elections.

The development and spread of democratic processes as a means of organising human society was a major evolutionary advance. It made it more difficult for narrow but powerful interests to bias the market and government in their favour. As a result, democracy has produced societies that are better governed and more competitive. And it has ensured better management by aligning the interests of government more closely with those of the society.

As we have seen, in the absence of democratic processes continual competition between societies is essential if the interests of rulers are to be continually and strongly aligned with those of the society. The competition has to be so strong and incessant that it provides immediate feedback to the ruler on the effectiveness of his management. But as human societies have grown larger in scale and fewer in number, competition between societies rarely provides continual and specific feedback of this strength. Democratic processes go some way toward ensuring that the interests of rulers and the society are aligned. Democracy produces governance that is closer to the ideal of ensuring that the benefits captured by members of society reflect their contribution to society.

Of course, in democratic societies as in any other human society, it will be in the interests of the economically powerful to do what they can to get governments to advance the interests of the wealthy. The economically powerful will benefit enormously if they can find ways to do this. But a government in a democratic society can serve the interests of the wealthy only if it can still maintain the support of a majority. If this is to occur, the majority must be convinced that a government that serves the interests of the wealthy is in fact serving the interests of the majority. To achieve this, the wealthy must influence the ideas and beliefs used by the majority to build their mental models of how the society functions. The mental models of the majority must lead them to conclude that the actions of such a government is in their interests.

On the face of it, this is no easy task. But the economically powerful are in a special position to be able to influence the ideas and beliefs that members of a society use in their mental models. Their wealth gives them control over most of the resources of the society. Most of the newspapers, television stations, and other media are owned by wealthy individuals and corporations. The economically powerful are major funders of universities and research institutes, and can determine the reputation of a university by deciding the employability of its graduates. Most of the members of modern societies are dependent directly or indirectly on the economically powerful for continued employment. Individuals who have and promote radically different views on how society should be organised have always had difficulty in obtaining well-paid employment in democratic as well as communist countries. IBM does not employ anybody in a senior position who uses his 'personal' time to campaign actively against multi-national companies, warning how they corrupt governments and undermine free markets. And political parties must obtain substantial financial contributions from wealthy individuals and corporations if they are to organise the large-scale election campaigns that are essential for electoral success in large modern nation states. These essential contributions will dry up if a political party promotes ideas and beliefs that offend the economically powerful[12].

The use of these opportunities to influence the nature of the ideas and beliefs that are treated as respectable in a society does not require any coordinated conspiracy on the part of the economically powerful. Any detailed cooperation amongst the wealthy as a group is undermined by free riding and cheating just as it is within any other large group of self-interested individuals. But the wealthy have a common interest in defending the enormously unequal distribution of wealth that has enabled them to accumulate and retain their riches. And they can exercise a substantial degree of influence without any significant self-sacrifice or investment of resources. It requires little extra investment for the wealthy to use their existing economic power to influence public debate. In the ways described, the economically powerful can determine the ideas and beliefs that are repeated and treated seriously in the mass media, that attract research funding, that are supported by expert opinion from universities and research institutes, are adopted by political parties, and are publicly supported by citizens who wish to remain in well-paid employment. And they can determine which ideas and beliefs will be quickly dismissed, ridiculed and, if necessary, demonised[13].

This manipulation of public ideas and beliefs is able to take advantage of the fact that most members of modern societies have not yet developed an effective capacity for systemic modelling. They cannot form complex mental models that enable them to predict how their society functions and how it would change under different conditions. Most members of modern societies have a limited capacity to evaluate for themselves even the simple models that are continually repeated and promoted in the public media. They have even less ability to develop for themselves more complex and more accurate models of how their economic system operates and how it could be improved. The simple economic models that are promoted through the public media serve the same function as the religious and moral beliefs that were used by rulers in earlier times to retain the support of their citizens. Just like a belief that the ruler is a god, or is chosen by the gods, the simple models allow the governors to pay less regard to the interests of the governed.

The economically powerful will ensure that simple models that advance their interests are repeated publicly and treated as authoritative. Typically these models show by simple chains of cause and effect that actions which favour those with economic power also help the less wealthy. They show that actions that would disadvantage the wealthy would also disadvantage the majority. In most democratic nation states at the turn of the 20th century, citizens are continually presented with simple models of these kinds. They are told that the way to reduce unemployment and to increase the living standards of the majority is to reduce minimum wages and working conditions, decrease social security benefits, and provide incentives to business through subsidies and tax concessions.

Citizens are continually shown how globalisation has made it impossible to collect much tax off large corporations and the wealthy without damaging the economy. This would cause a flight of capital that would increase unemployment and reduce living standards. They are told that countries cannot now afford the welfare and social security systems that in the 20th century have prevented the misery and the poverty that unmanaged capitalism produced for many in the 19th century. As the need arises, simple models are rapidly developed and promoted to show that anything done to directly help the poor will in fact harm them, the only way to help the poor is to help the rich, and anything that harms the rich will hurt the poor. And whatever crisis emerges in the society and its economic system, whatever its cause, simple models are soon developed to show that the crisis arose because workers' wages are too high, or because welfare schemes are too generous.

It is not difficult to develop simple models of this type that have a surface plausibility. In a complex society most actions have a multitude of effects, some of which will advantage a particular group, some of which will not. Any action that advantages the wealthy will probably also have some effects that advantage the poor, as well as some that harm the poor. It is a simple matter to construct models that highlight only the causal chains that suit a particular argument. Such a model will not be wrong because of what it contains, but because of what it leaves out. The logical and causal connections used by the model will be valid. But minds that cannot model the complexity that has been left out of the simple model will not be able to see that the model is inadequate and manipulative.

* * * * *

Looking back at the major issues discussed in the last two Chapters, we can see that the ability of human societies to exploit the benefits of cooperative organisation has improved considerably in the 10,000 years since external management first emerged. The size of human societies has increased enormously, establishing cooperative organisation over much larger scales. Improvements in the ability of rulers and governments to use systemic modelling has enhanced the evolvability and competence of their management. They now do not have to rely as much on the existence of distributed internal management in the form of moral and religious belief systems. This has further enhanced evolvability: societies with fewer members who are straight-jacketed by inculcated beliefs are better able to adapt and explore new possibilities.

These improvements in management enabled governments to establish and adapt the framework of controls that allow large-scale economic markets to emerge. Economic markets have an extraordinary ability to exploit the benefits of cooperation. They produce cooperation because participants are able to capture the beneficial effects on others of their actions. In a market, cooperation pays. The ability of markets to organise cooperation has enabled them to produce the complex and efficient cooperative division of labour that characterises modern nation states.

Where economic markets work effectively, they are far superior to governments at organising cooperative economic activity. But markets are not able to exploit the benefits of cooperation fully in all circumstances. This is the case for cooperation that has collective effects. Markets are unable to ensure that citizens capture all the effects of actions that have collective effects. So ungoverned markets fail to reward and organise activities that have collective social benefits such as defence, education, and programs that build better communities. And they fail to deter pollution and other activities that cause collective harm.

Markets also fail to develop fully the potential of all members of a society. This is because markets produce a massively unequal distribution of wealth, and then they ignore the educational and other needs of those they leave with little purchasing power. Because income and purchasing power is concentrated in a few hands, markets organise the society to mostly serve the interests of a small minority.

By concentrating wealth, markets also produce a small number of wealthy individuals and corporations who have the power to get governments to bias the market and other government actions in their favour. Democracy has only limited success in countering this manipulation of governments.

So market systems do not make governments redundant. Management by governments is essential to establish and adapt the market framework and to correct the deficiencies in the market system. Where markets fail, government action is necessary to ensure participants capture the full effects on others of their actions, and to ensure participants do not capture greater benefits than they should. Government can use taxes, laws, subsidies and other payments to try to organise the behaviour and activities that would have arisen if the market was not deficient. They can correct market failures by ensuring that citizens capture all their effects on society.

But the forms of government that are currently in existence are seriously limited in their ability to carry out these functions effectively. Governments will never have the detailed information needed to determine accurately how the needs and values of the members of the society are best satisfied. In a large and complex society, governments also have to rely on incompetent and inefficient bureaucracies to plan and implement their management. And the interests of government will not be fully aligned with those of the society. External competition will rarely be strong enough to do this, and democracy has been unable to prevent government from being manipulated in the interests of the economically powerful.

As we shall consider in detail in the next Chapter, humanity must find better ways to build evolvable cooperative societies if we are to participate successfully in the future evolution of life in the universe.

17

Planetary Society and Beyond

How must human society evolve if we are to participate successfully in the future evolution of life in the universe? How should we change the way human society is currently organised if we want future evolutionary success?

We have seen what we must do as individuals. We must develop the psychological capacity to adapt in whatever ways are necessary to achieve future evolutionary success. This means that we must no longer pursue motivations and values set by our biological and social past. Instead, we must develop the psychological ability to find motivation and emotional satisfaction in whatever we need to do to achieve future evolutionary success. We must use mental models of our evolutionary future to set our ultimate objectives, and then align our emotional system and other pre-existing adaptive processes with these evolutionary objectives.

But we must also organise ourselves socially in whatever ways will enable us to best exploit the benefits of cooperation. Whatever it is that we have to do for evolutionary success in the future, cooperative organisation will enable us to do it more effectively. We have seen that during the past evolution of life on earth, the larger the scale of cooperative organisation and the more evolvable it is, the better it is at exploiting the benefits of cooperation, and the better it does in evolutionary terms.

We can expect that the same principles will apply to future evolution.

The larger the scale of human cooperative organisation, the larger the scale on which humanity can organise matter and energy to adapt to whatever challenges confront us. The greater the evolvability of human organisation, the better it will be at discovering and supporting the most effective cooperation between its members. And the better it will be at discovering new adaptation, including technological and scientific advances, and at adapting as circumstances change.

Of the instances of life that emerge throughout the universe, the ones that will be most significant will be those that progressively evolve cooperative organisation of larger scale and greater evolvability. When life evolves successfully on a planet, it will eventually produce a cooperative organisation on the scale of the planet. The planetary organisation will be established by external management that supports cooperation and suppresses destructive competition within the organisation. The members of the planetary organisation will be aware of the direction of evolution, and will consciously design and structure their social organisation so that it pursues evolutionary objectives. They will consciously improve the evolvability of their planetary organisation, structuring it so that it can adapt for the outside/future as well as for the inside/now. Like each of its member organisms, the planetary organisation will be able to adapt as a coherent whole, pursuing evolutionary objectives by acting on its external environment, and able to relate to any other large-scale living organisations that it encounters.

The instances of life that will be most significant in future evolution will be those that continue to expand the scale of their cooperative organisation. As an organisation expands, it will be able to manage the matter and energy of a solar system, then of a number of stars, and then of a galaxy. The range of things it can do in response to adaptive challenges will increase substantially as it gains the ability to manage matter and energy over wider and wider scales. As it expands, an organisation may amalgamate with cooperative organisations of other living processes, forming a larger-scale organisation containing a diversity of instances of life. Organisations will become expert at forming large-scale cooperative organisations that preserve and advance the interests of their members, and that exploit and celebrate the diverse perspectives of their members for the benefit of all. The members will be united by the common objective of contributing to the successful evolution of life in the universe, wherever that takes them.

The direction in which humanity must head for future evolutionary success is clear. The first great challenge is to establish a highly evolvable cooperative organisation on the scale of the planet. Humanity is likely to move some way toward such a planetary society even if we do not consciously pursue evolutionary objectives. This is because planetary cooperation will provide immediate material and social benefits to humans. These benefits have already driven the expansion of economic markets. Markets now operate globally, organising cooperative divisions of labour on the scale of the planet. Markets have increased in scale because the actions that cause them to expand are in the interests of market participants. Markets that are wider in scale cover a greater diversity of participants and provide more opportunities for useful specialisation and division of labour. Individuals who expand the scale of markets by exploiting these opportunities benefit from doing so[1].

But the processes of government do not yet operate on the scale of the planet. The scale of cooperative human society organised by external management has expanded enormously since the first agricultural communities emerged about 10,000 years ago. We now have nation states that include hundreds of millions of people. Yet this falls far short of a system of governance that would span the globe and would manage all humanity into a unified but diverse cooperative organisation.

The scale of governance is likely to continue to increase. A system of global governance is in the interests of humanity whether or not we consciously adopt evolutionary objectives. The deficiencies of the market system and the immediate material and social benefits of wider-scale cooperation will drive the move toward global governance. We have seen that a cooperative and effective society cannot be produced by a market system alone. This is as true for global society as it is for any other society. There are a number of reasons why global governance is in the immediate interests of humanity. First, an effective market can emerge only if there is a system of governance to establish the framework of market controls. If participants can escape these controls, they can cheat or free ride, undermining the market. If global markets are to be effective, all participants must be subject to a system of global controls that can be adapted swiftly and efficiently as circumstances change. To be effective, global markets demand global governance.

Second, global markets share the deficiencies of all other markets. They are unable to establish economic and social activities that have collective benefits, prevent environmental degradation and other collective harms, or produce a distribution of income that ensures all members of society can realise their full potential as individuals and collectively. A global society must be managed by global governance if these limitations are to be overcome, and if the benefits of cooperation on a global scale are to be realised. The governments of nation states can attempt to correct these deficiencies within their borders, but cannot succeed where the causes of the deficiencies extend beyond the borders of any one state. Only a larger-scale government that can manage the behaviour of all nations and their members can do this[2].

Global governance is essential if nations are to cooperate to overcome the deficiencies of global markets. Global governance can ensure that nations capture any harms they or their members cause by degrading the global environment. It can ensure that nations capture the benefits of their contribution to global social, economic and environmental projects that have collective benefits across nations[3]. Only global governance has the power to redistribute wealth between nations in the same way that nations do between their members. Only global governance can prevent war and other destructive competition between nations by using its power to ensure that nations capture the harms they produce[4]. And global governance is the most effective way of ensuring that all nation states use systems of governance that are fair and develop the potential of all their citizens[5].

Finally, global governance is in the immediate interests of humanity because global markets are undermining the ability of nation states to govern in the interests of their citizens[6]. We have seen that nation states must be governed effectively if the deficiencies of their internal markets are to be corrected, and if they are to fully exploit the benefits of cooperation within the state. But global markets are increasingly unravelling the systems of laws, taxes and government programs that nations have put in place. This is because these systems add to the costs of producing goods and services, and producers in countries with lower costs will be more competitive in global markets. Governments price their producers out of the market if, for example, they maintain higher minimum wages than other countries[7], or stronger workplace safety standards, or more effective pollution controls, or if they require employers to contribute more to employee pension funds, or to pay higher taxes so that government can fund a better social security system and a more effective education system. When producers sell goods and services in international markets, they are unable to capture any of the effects of providing these benefits to their employees or to others in their country. The market does not give any additional returns to producers who provide these benefits. In a competitive global market, producers cannot set prices that recoup their contribution to these types of benefits.

Producers who are able to manufacture the same goods anywhere in the world will locate in a country where they can do so most cheaply. Those who do not will be uncompetitive in global markets. Governments have to reduce the cost to business of government regulation and taxation if the nation's economy is to be competitive. Governments that do not do so will be undercut by others. Competition will drive down the costs that governments can impose on business, and therefore will undermine the ability of governments to correct the deficiencies of the market. This will be the case no matter how beneficial the government regulation would have been to the society. Gradually, competition will dismantle the social security systems and other protections that governments have introduced this century to make the market system bearable[8]. Governments will not be able to fund these programs by increasing taxes on the income and expenditure of the affluent. The wealthy will use their political power and their increasing mobility between countries to ensure that any attempts in this direction fail.

Those nations that enter global competition with an initial competitive advantage will be able to maintain these government protections for longer. For example, a nation might have a relatively better education system and a more highly skilled workforce. This might give the nation a competitive superiority that enables it to resist the forces that are undermining the governance of nation states. But it is very unlikely that any country can maintain a competitive advantage for any length of time that is sufficient to continue to fund the programs of a welfare state. Other countries will imitate the successful one, and eliminate its competitive advantage. The dismantling of social security systems is proceeding apace in even the world's most powerful economies.

The only actions of governments that can survive global competition will be those that make their producers more competitive in the market. All other management activities, including those that are needed to correct the deficiencies of the market, will be undermined. Unchecked, this would eventually return us to the unmanaged markets of the nineteenth century, and the widespread misery, poverty and inequality that they produced[9].

The only way in which this process can be stopped is by some form of global governance. Individuals, corporations and nations are powerless to prevent the disintegration of society that unmanaged markets will produce. Even if nobody in the world wants it, unmanaged global markets will continue to undermine national government. Anything a nation does to counter the process will simply make business within the country less competitive. Only global governance can end the competition between national governments that is eroding their ability to correct the deficiencies of the market. Global governance is essential to establish and enforce minimum global standards for wages, working hours, and social security. Only the power of global governance can prevent countries from undercutting these minimum standards and continuing the downward spiral into environmental and social disintegration that is being produced by unmanaged global competition. Global governance can establish minimum standards of living to provide a floor from which business competes in all countries of the world.

Just as the misery produced by unmanaged markets in the nineteenth century made the modern welfare state inevitable, the misery produced by unmanaged global markets will make global governance inevitable.

The only unanswered question is how much misery will be produced before global governance is established. Effective global governance is in the interests of the majority of people in the world. If they had a capacity for systemic mental modelling that enabled them to see and understand this, global governance would be established swiftly. The majority would use democratic processes to have their national governments get together with other governments to set up such a system. To an extent, the European Union is a small-scale example of how this could happen[10].

But the majority of people on the planet do not yet have the knowledge or the ability to form mental models of how global markets operate and how various alternative forms of global and national governance would interact with those market systems. And the economically powerful are using their influence over public ideas to reinforce simpler and less adequate models. These models point to the evils rather that the benefits of national and global governance. It is in the immediate interests of the wealthy to support the dismantling of the social security and other programs of national governments. They do not need these programs. Their wealth enables them to satisfy their own needs, and to avoid the effects of social disintegration. But governments currently make them contribute to the funding of the programs through their taxes. The wealthy have the most to lose from attempts by governments to redistribute income and to correct other failings of markets. In many industrialised countries, the wealthy have pursued their interests by using the harmful effects of global competition to gain support for dismantling the welfare state even more quickly than the market would have required.

In the long run, however, the economically powerful will see that effective global governance is as essential to their interests as were effective systems of national governance earlier this century. Large-scale social unrest is as inevitably a consequence of unmanaged global markets as it was of the unmanaged market systems of the nineteenth century. And this social disintegration will eventually threaten the interests of the economically powerful, as it did earlier this century. Eventually, it will be clear to all that global governance is essential to correct the disadvantages of the global market.

Effective and adaptable global governance can ensure that the enormous advantages of global markets can be enjoyed without the disadvantages. Global markets can go a long way toward ensuring that citizens of the planetary society will capture the effects on others of their actions. To the extent that global markets achieve this, they make it in the interests of citizens to cooperate to satisfy the needs of others. Where markets fail to do this, government intervention is essential to organise society cooperatively. Governance at the level of the neighbourhood, city, region, nation, and the planet can act to ensure that members of the planetary society capture all the effects of their actions, good or bad, on others. To the extent that governance achieves this, human society will fully exploit the benefits of cooperation. Where cooperation can produce overall benefits, effective governance can ensure that individuals will find it in their interests to act cooperatively.

But, as we have seen, our existing systems of governance will not do this efficiently and effectively. Our existing forms of government are incompetent. Where governments have competed directly with markets in the organisation of economic activity, governments have proved inadequate. The same deficiencies in government that make it inferior to the market also limit its ability to organise the features of society that markets cannot. As we have seen, governments lack the information they need to manage a complex society effectively. They also must rely on incompetent bureaucracies to design and implement their policies. And their interests are often not aligned with the interests of the society as a whole.

Management is essential to produce an effective, cooperative planetary society. But our current forms of government are seriously limited in their ability to discover and adapt the management that is needed. If humanity is to fulfil its evolutionary potential, it must discover new and better processes for establishing the various levels of governance. The development of a managed planetary society is not enough. The next great evolutionary challenge for humanity will be to invent a new form of governance that will overcome the serious flaws in our existing processes.

We need a system that is as superior to current governance as is the market. We need a system for establishing and evolving governance that is as dynamic and evolvable as the market. It must be superior at discovering better forms of governance, and at adapting them as circumstances change. Highly evolvable management is essential if we are to have a highly evolvable human society.

What features would a system of governance have to have if it were to match the evolvability of the market? As we have seen, the market owes its effectiveness to a number of key features. Markets enable individuals to capture the effects on others of their actions. Individuals capture these effects when they sell the goods and services they produce. This aligns the interests of producers and consumers, and ensures that the energy and creativity of the members of society are harnessed to satisfy the needs and wants of others. There are no restrictions on who can try to produce and sell particular goods and services. This generates competition that ensures that producers survive only if they satisfy the needs of consumers efficiently.

As we have seen, the market mechanism uses a supra-individual change-and-test process to search for better ways to satisfy the needs of the members of a society. The change-and-test process operates through the collective activities of market participants. The market tries out new goods and services by providing incentives for producers to develop and market them. Producers, like central planners, can never be certain which products will satisfy consumers best. So the market uses the purchasing decisions of consumers to test the 'best guesses' made by producers. In a market, it is consumers rather than incompetent bureaucracies or governments who decide whether particular goods and services are useful to consumers, and how much they are worth.

Taken together, these collective actions of producers and consumers operate a supra-individual change-and-test mechanism that can discover better goods and services. The mechanism can do this even though no individual within the market knows which products will prove to be better. As a result, the market is more effective at matching goods and services to the needs of citizens than any group of central planners, or any producer acting alone. And the entire process operates through individuals pursuing only their own immediate interests.

Is it possible to establish similar processes to produce and adapt the functions currently performed by government? Could governance be established by supra-individual change-and-test processes that continually search for and implement better governance? Could these processes replace our current political processes and systems of government?

Such a system would have to operate a market in the various things that are currently done by governments. It would have to establish what I will call a vertical market, to distinguish it from our current economic markets. In a vertical market, it would not be the goods and services traded in economic markets that would be produced and sold. Instead, the vertical market would trade in regulations, market frameworks, systems of education, laws, taxes, law enforcement systems, and programs that build better communities. Any component of governance could be developed and sold in a vertical market. The 'goods' of the vertical market would be any act of governance that could organise some part of the society more effectively[11].

Like goods in economic markets, acts of management and components of governance would be purchased by the members of society who benefit from them. As we have seen, effective governance can produce net benefits for citizens because it enables the advantages of cooperation to be exploited. Governance can do this by supporting beneficial cooperative activities directly (e.g. by buying a defence force), or by creating a framework of market controls that ensure individuals can pursue their own interests only by cooperating. So governance can provide net benefits for members of society by, for example, redistributing income, improving community life, preventing environmental degradation, enabling economic markets to emerge or providing general education. Acts of governance that produce net benefits in these ways could all be sold at a profit in a vertical market.

Unlike in economic markets, the products sold in a vertical market would not be purchased by individuals. Acts of governance commonly affect many people, not just individuals. So in a vertical market, acts of management would not be purchased by individuals. They would be purchased collectively by the group of citizens who benefit from the management.

Vertical markets would be fully open to competition. No one would have a monopoly on developing the various acts of management that are put in place. Unlike at present, there would be no government with a sole right to decide the governance that is implemented. Any individual or corporation could develop an act of management and attempt to market it. If the act produced superior benefits, it would be purchased and implemented. Competition between producers of acts of governance would ensure that governance was produced efficiently.

The result would be a system that used a supra-individual change-and-test process to determine how a society is governed. Entrepreneurs would develop alternative acts of management and attempt to sell them to those who would benefit from them. The acts of governance that were established would be those that could be sold at a profit—those that could produce net benefits to members of the society. By making good governance profitable, the vertical market would give entrepreneurs an incentive to develop ways to improve governance. And like the horizontal economic market, the effectiveness of the collective change-and-test processes of the vertical market would not depend on entrepreneurs getting it right. Any particular act of management that was purchased by consumers might have been developed incompetently. It might have hit the mark by chance, like a random mutation that happens to work. Entrepreneurs would not have to be any better than central planners for the vertical market to produce better governance. A process that uses the best judgements of entrepreneurs in a change-and-test mechanism can be far superior to a process that relies on the judgements of entrepreneurs (or central planners) alone.

This completes a brief outline of a vertical market system. But before we move on to a more detailed description of how it would operate, we need to look in greater depth at the key condition that must be met if a vertical market is to emerge. A vertical market can operate only if entrepreneurs are able to obtain appropriate payment from all those who benefit from the acts of governance that the entrepreneurs develop. Then, if the total of the benefits produced by an act of governance exceed its costs, it will be profitable for an entrepreneur to develop and implement the act. If entrepreneurs are able to make a profit by designing and establishing better governance, a vertical market can emerge. But a vertical market will fail if all those who benefit do not pay for the benefits they receive.

So why is it that entrepreneurs do not develop and market better governance at present? If, like better goods and services, better management can produce net benefits to others, why are entrepreneurs unable to make a profit out of improving governance at present? The reason is the free rider problem. Even though an act of management might be in the interests of all members of a society, it is in the interests of individuals to let someone else pay for it. If the taxes needed to provide effective defence or a system of public roads were voluntary, many would not pay. As we have seen in detail, governments and rulers can overcome this problem by using their power to prevent free riding. They can force everyone to pay for collective benefits by making tax compulsory.

An entrepreneur can profit from establishing an act of governance only if all who benefit pay for the governance. A vertical market can operate effectively only if the free rider problem is overcome. Just as economic markets can emerge only if a framework of controls exists to prevent cheating and theft, a vertical market needs a framework of controls that prevents free riding. The simplest way of overcoming this problem would be a requirement that all who benefit from an act of governance must pay for it if the majority decide to buy the governance. If such a requirement applied, it would be in the interests of all who benefit sufficiently from an act of governance to support a decision to purchase it. There would no longer be any way in which they could get the benefits without paying for them. Free riding would no longer pay.

The framework of controls for the vertical market would also have to ensure that any proposed acts of governance would fully compensate for any harmful effects they caused. This would ensure that acts of governance capture their full effects on members of society. Of course, the exceptions to this are where an act of governance would harm individuals to ensure they capture the effects on others of their harmful acts, or where governance would prevent individuals from capturing greater benefits than reflects their effects on society. So acts of governance that prevent cheating and free riding, or that redistribute income away from individuals who earn high incomes because they are sheltered from competition, would not have to compensate the individuals they disadvantage in these ways.

It is likely that a vertical market would redistribute incomes produced in the vertical and horizontal markets by establishing a minimum income system for all members of planetary society. This would provide a minimum standard of living for all, and guarantee that everyone received an education and the other resources they needed to develop their potential to contribute to society. It would also ensure that citizens benefited sufficiently from their participation in society to attract their support for the way the society was organised. And it would guarantee that everyone had sufficient purchasing power in the vertical market. The extent of the redistribution, the level of the guaranteed minimum income, and conditions attaching to its use would all be decided in the vertical market. The market would balance the negative and positive effects of various aspects of the system, including the impact of the redistribution of income on the incentives for participants in the economic and vertical markets.

The framework of controls that establish the vertical market would also be established, adapted and improved through the vertical market. An entrepreneur could propose particular changes to any aspect of the framework, and attempt to sell them to all who would be affected. It would be profitable for an entrepreneur to develop more efficient and effective ways of preventing free riding, cheating, and theft, or to appropriately redistribute income earned in the vertical market, or to open areas of the vertical market to greater competition. Under a vertical market everything would be contestable, including all the elements of the vertical market itself.

Once a vertical market was established, its framework as well as other acts of governance would be free to evolve and improve. Changes that produced net benefits would be implemented. As a result, a vertical market would be self-repairing and self-improving. If any aspect of the vertical market were found not to be operating efficiently and effectively, there would be a profit in finding a way to fix it. If the governance of the society failed in some respect to ensure that the benefits of cooperation were being fully exploited, there would be a profit in new acts of governance that rectified this. And if this failure was due to a flaw in the vertical market itself, there would be a profit in adapting the market framework to correct the failure. The vertical market would self-organise to ensure, as far as possible, that all members of society captured the effects on others of their acts.

For example, a newly implemented vertical market might initially be limited in its ability to establish some acts of governance that are very complex and have complicated effects. The consumers who would benefit from them might not understand how the governance would operate, and be unable to evaluate alternatives. In these circumstances, initiatives that overcame this difficulty could be profitable in the vertical market. For example, it might be in the interests of the entrepreneurs who develop acts of governance to put more resources into marketing and into educating consumers. It might also be profitable for other entrepreneurs to develop independent educational material, or to establish independent teams of experts to evaluate and advise consumers on the benefits of particular proposals. It might be profitable for entrepreneurs who develop new complex acts of governance to be able to show to consumers that they have submitted themselves and their proposals to evaluation by independent experts. Or, in some circumstances, consumers might prefer to delegate the decision about which governance to buy to other bodies. These bodies could be organised and staffed so that they make competent decisions and so that their interests are aligned with the interests of the consumers.

Just as it is impossible to predict the specific goods and services that will be developed in the economic market in the future, it is equally impossible to do so for the vertical market. It is not possible to foresee the specific acts of governance, including those that form the framework of the system itself, that will be developed in a vertical market. As has been the case with economic markets, the innovation and creativity unleashed by a vertical market would produce governance of a diversity and variety that would be inconceivable to those who lived before the market was established. We can be sure that vertical markets would get better and better at managing society to satisfy the needs and aspirations of its members, but the exact details of how it would achieve this is unknowable.

Of course, the vertical market would establish whatever form of governance the members of society wanted to purchase. In theory, it would be open to a planetary society to use a vertical market to purchase a democratic political system along the lines we have at present. If the members of the society could not envisage a better system, and if entrepreneurs could not develop and sell refinements and improvements, a democratic political system could continue indefinitely within a vertical market system. The dead hand of government as we know it could live forever. But this is highly unlikely.

It is much more likely that a vertical market would eventually produce a highly layered, differentiated, diverse and dynamic form of governance. The ability of vertical markets to facilitate the differentiation of governance would be similar to the capacity of horizontal markets to encourage the differentiation of products and services to satisfy every taste and preference. Whenever entrepreneurs developed and sold any new act of governance that benefited a section of society, the overall system of governance would increase in diversity. The opportunities for this diversification and specialisation would be enormous. Groups such as the tenants of a block of flats, a neighbourhood, a small rural community, an industry, a school community, a town, a city, and a region might all discover specialised acts of governance that manage their members and produce benefits for the group. The vertical market would enable them to explore and exploit these opportunities.

These groups would develop their governance within a framework of wider-scale governance established by the vertical market. This wider-scale management would ensure that the ability of smaller-scale groups to establish governance was not abused. It would guarantee that governance could be erected only if it met the test of ensuring that individuals captured the effects of their actions on others. It would therefore not enable any individuals to persecute others within the group or to obtain more benefits than were necessary to reflect their contribution to the group. And these higher layers of governance would also ensure that groups captured the effects of their action on other groups and on individuals outside the group. The highest layer of governance would operate on the scale of the planet, ensuring that individuals and groups captured the effects of actions that had global consequences.

The vertical market would have the capacity to establish new markets and to modify existing ones by changing their regulatory frameworks. It would decide between alternative methods of governing, including between markets and central planning. We saw earlier that in some circumstances the best forms of governance are those that establish markets that include supra-individual change-and-test processes. This can be far more effective than governance that centrally plans and implements detailed programs that support specific cooperative activities. Markets that include change-and-test processes are particularly effective when there is insufficient information or knowledge available to design an optimal solution to a problem of governance. In these circumstances, the best that might be done is to establish a specific market process that searches for the best solutions by combining the intelligence of participants with trial-and-error.

The vertical market would decide when and where market processes rather than specific programs were used to organise cooperation. New markets would be established and existing markets modified by purchases made in the vertical market. The vertical market would determine the content and structure of the regulatory frameworks that enable economic and other markets to operate. For example, the vertical market could consider proposals to create a market in rights to produce carbon dioxide emissions[12], or for an 'ideas futures' market (ideas futures markets are processes that enable the strength of expert opinion on controversial issues to be gauged accurately. They are particularly effective at ensuring that expert opinion is not biased by conflicts of interest[13]). The vertical market would be a meta market—it would operate markets in markets

One of the greatest strengths of a vertical market system is that it would align the interests of those who produce governance with the interests of the governed. As in an economic market, consumers would ultimately control what is produced. The only way a producer of governance could pursue his interests would be by selling governance to those affected by it. And competition between producers would ensure that they could prosper only by producing efficient governance. In a competitive vertical market, the only way an entrepreneur could advance his interests would be by developing management that served the interests of those who would be affected by it. The producers of governance would capture the effects on others of their actions. As a result, the economically powerful would have far fewer opportunities to bias and manipulate governance than under our current systems. And whenever the economically powerful found a new way to do so, there would be a profit in the development of new acts of governance that ended the bias. Vertical markets would continually search for ways of ensuring that economic power is not abused.

But there is another major advantage to this alignment of interests between those who produce governance and those who are governed. A vertical market would enable any member of society to capture the benefits of anything he could do to improve an aspect of the governance of society. As a result, the vertical market would provide incentives for members of society to continually search for better governance. Economic markets harness the motivations and creativity of members of society to continually develop and produce better goods and services. In the same way, a vertical market would harness the energies and abilities of members of society to improve the various functions that governments now perform.

Just as an economic market makes it profitable to develop a more fuel-efficient motor car engine or a tastier ice cream, a vertical system would make it profitable to develop a better regulatory framework for an economic market, or a program that enhances the quality of life in a particular community, or initiatives that reduce the perverse incentives of a welfare scheme. A vertical market would make it profitable to develop governance that better ensures that polluters pay the full cost of the harm they cause, and governance that provides all members of a society with the material and educational resources needed to realise their potential to contribute to the society. It would make it profitable to organise initiatives that cure the sick and feed the starving.

The energy, drive and creativity that the economic market harnesses to produce better perfumes, mouse traps and hair dryers would also be harnessed to produce better ways of undertaking the functions of government. Where entrepreneurs believed there was significant potential for improvements in particular aspects of governance, they could profitably organise the finance to put together an expert team to develop and market the improvements. Just as corporations are formed to exploit the potential profits to be made from developing better goods and services, they would also be formed to exploit the profitability of improving governance.

Any aspect of governance that did not ensure that members of society captured the effects on others of their actions would provide profitable opportunities for anyone who could devise better governance. There would be advantage in finding an uncaptured harm or benefit, and ensuring that it was captured. Greedy and ambitious young men would dream as much about finding an uncaptured harm or benefit as they would about being the first to develop a better good or service for the economic market. The dead hand of bureaucracy and government would be replaced with the energy and creativity of the market.

As in economic markets, the consumers of governance, not central planners, would decide which acts of governance were better for consumers. The effects of governance would be evaluated by those who were in the best position to do so: by those who felt its effects. A vertical market would provide entrepreneurs with the incentive to use human knowledge and mental modelling to develop acts of governance designed to best advance the interests of the governed. But the effectiveness of the vertical market would not depend on entrepreneurs designing the best forms of governance. The vertical market recognises that due to a lack of relevant information and the complexity of circumstances, entrepreneurs, like central planners, would often get it right only by luck. And entrepreneurs could never know accurately when they had chanced upon the right answer. To overcome this difficulty, the vertical market uses a supra-individual change-and-test process to discover the best forms of governance. It complements the knowledge and mental modelling of entrepreneurs with trial-and-error processes in which the consumers of governance evaluate the trials. The human mind and mental modelling has not been able to replace the internal change-and-test processes that adapt our bodies physiologically. Equally, they cannot remove the need for supra-individual change-and-test processes to adapt our societies.

Of course, democratic political processes also operate a supra-individual change-and-test process in which changes to governance are evaluated and tested by citizens. But democracy is far inferior to a vertical market at discovering better forms of governance. In a typical western democratic system, a small number of political parties each develop a package of governance that they put to the people at elections. Often only two of the parties have a real chance of forming a government. By electing a government, the people choose much of the governance that will apply until the next election, usually about 3 to 5 years away.

To see how ineffective such a democratic system is at matching governance to the needs of the governed, it is useful to consider how ineffective an economic market would be if it were organised in a similar way. A comparable economic market would be one in which there were only two possible producers of goods and services. Each possible producer would develop a package of all the goods and services that each citizen would have over a three or four year period. Citizens would then choose which package would actually be implemented by voting for the producer of their choice. They could not pick and choose goods and services out of each package. They could have only one package, in its entirety. Such a system would obviously be far inferior to our current economic markets. Our present democratic systems of governance share all the features that would make such an economic system incompetent at satisfying the needs of consumers.

Vertical markets would also be far superior to citizen-initiated referenda at discovering and adapting effective governance. Referenda are a way of trying to overcome some of the limitations of a system in which citizens vote only every three or four years, and must accept or reject packages of policies as a whole. Referenda enable citizens to vote on specific issues more or less continuously. But a fundamental limitation of current referenda systems is that they do not include processes that properly fund the development of the proposals that are put to the vote. In contrast, the vertical market ensures that the development of proposals for better governance is profitable. Where it leads to better governance, a vertical market makes it profitable to use substantial resources to undertake research, or to develop complex models to evaluate alternatives, or to employ creative and talented people to assist in the development of a proposal. It does this in the same way that economic markets fund the development of new goods and services.

Referenda are good systems for testing proposals, but not for generating them. As a result, their ability to discover new and better forms of governance is very limited. They are as ineffective as would be an economic market that relied largely on consumers to design and invent goods and services. This is not the only serious limitation to the evolvability of existing systems of citizen-initiated referenda: current systems do not allow referenda to change the framework of rules that establishes the system. The systems themselves are not very evolvable.

A vertical market system would have clear advantages over our current systems for establishing governance. It would be better at discovering management that is more effective and efficient at exploiting the benefits of cooperation. It would produce a more cooperative and evolvable planetary organisation. The establishment of a vertical market would be as significant a step in the evolution of human society as was the establishment of economic markets to replace much of the central planning by rulers.

It would be in the interests of the majority of members of human society to put an end to government and politics and to replace them with a vertical market system. But the majority will see this as in their interests only once they have developed a capacity to mentally model vertical markets and alternative systems of governance. And those who do better under current governance will do what they can to undermine the development of this capacity as well as oppose any move to establish a vertical market. An effective vertical market will ensure that citizens capture the effects of their actions on others in the society. Citizens will do no better than this, nor worse. But, as we have seen, under our current system of governance a small minority is able to accumulate and maintain great wealth by obtaining benefits that are way out of proportion to their contributions. The economically powerful have a lot to lose in any move toward a vertical market. They can be expected to use the various opportunities they have to influence the members of society to oppose a vertical market. Of course, this will change as the economically powerful develop a capacity for evolutionary self-management. Once they are able to transcend their narrow biological and social needs, and instead pursue evolutionary objectives, they will support the establishment of a vertical market. Once they develop an evolutionary perspective, they will see that a life dedicated to the accumulation and use of wealth to satisfy their narrow biological and social needs would be a wasted and absurd life.

But a vertical system by itself will not produce a highly evolvable human society that uses the power of cooperation to pursue future evolutionary success. The introduction of a vertical market will not complete the ninth organisational milestone that began 10,000 years ago with the emergence of the first human societies governed by external managers. This is because a vertical market system does not establish the objectives of a society. Instead, a vertical market organises cooperation to satisfy the objectives and preferences of the members of the society. It does not matter what these objectives and preferences are. The vertical market will organise cooperation to pursue their satisfaction. If the objectives of most of the members of a society are to satisfy only their narrow biological and social needs and urges, the vertical market and the horizontal economic markets managed by it will satisfy only those needs. If pursuit of these needs will lead to future evolutionary irrelevance or death, the vertical market will pursue them nonetheless. The vertical market finds the best means of achieving whatever ends are set by the members of society. It ensures that governance serves the interests of the governed, whatever they may be.

So until the majority of members of a planetary society develop a capacity for evolutionary modelling and self-management, a vertical market would manage the society to satisfy only the narrow biological and social needs of its members. The members of society, their technology and science, and the resources of the society would all be managed by the vertical market to serve the narrow needs established in humans by limited and inferior evolutionary mechanisms. The result would be a planetary society that goes nowhere in evolutionary terms.

But once sufficient members of human society developed evolutionary consciousness, all this would change. The vertical market would be harnessed by the new values and objectives of these members to search for better ways of organising the society to pursue future evolutionary success for humanity. As humans developed objectives, motivations and values that reflected their improved ability to adapt for the outside/future, so to would the planetary organisation. As humans developed the capacity to take into account the consequences of their actions over greater scales of space and time, including over evolutionary scales, so to would the planetary organisation.

It is important to note that individual members of the planetary society would not be able to pursue evolutionary objectives effectively unless the society also pursued evolutionary objectives. Until the society pursued evolutionary success, the resources of the society would not be organised to support activities that furthered evolutionary objectives. Understanding this, those who achieved evolutionary consciousness would promote the development of evolutionary consciousness in others in the society, and would also support the establishment of an effective vertical system that could respond to their evolutionary objectives and values. Once sufficient members of a planetary society developed evolutionary consciousness, the vertical system would reward activities that contributed to evolutionary success. It would ensure that it was in the immediate interests of citizens to do the things that were necessary for future evolutionary success. It would ensure that individuals in the society captured all the effects of their actions on the long-term evolutionary success of the society. In effect, it would immediately feed back to individuals the longer-term consequences of their acts to ensure that they took these into account when deciding their actions.

The vertical system would organise humanity to manage all the resources of the planet for evolutionary objectives. The vertical system would manage the human members of the society and through them, their technology and science, other living organisms and processes on the planet, and the matter and energy of the planet to pursue evolutionary objectives. This would produce a planetary organisation that included the sum total of all processes on the planet, living and non-living, that could be managed to contribute to the future evolutionary success of the organisation. The organisation would not contain only humans. But humans would set the ultimate objectives and values that would be pursued by the vertical system and therefore by the planetary organisation as a whole. The vertical system would then determine how these objectives were best achieved. It would determine how humans, artificial intelligence, other technology, and other living processes would be organised and utilised to best serve these objectives. And it would adapt the ways in which these resources were used as new technology and living processes were developed, and as new evolutionary challenges were met.

In order to achieve continued evolutionary success, such a planetary organisation would have to organise itself to form internal models of its future evolution. It would have to develop the ability to use these models to discover organisational adaptations that would produce future success. It would have to develop the capacity to adapt coherently as a whole in the ways identified by its internal models. Almost certainly, it would also have to develop the ability to move, to expand its scale to that of the solar system and then to the galaxy and beyond, to remodel its physical environment, to have physical impacts on events outside itself, to form intentions, to establish projects and long-term objectives for the organisation, to communicate and interact with any other living processes that it encounters, to amalgamate with other societies of living processes to form larger-scale cooperative organisations, and to do any other thing that might produce success in the future.

To be assured of continued evolutionary success, a planetary organisation must become a self-evolving organism in its own right. Its ability to adapt and evolve must not be limited by the biological and social past of the living processes that participate in it. Through appropriate management of these living processes, the planetary organisation must be able to adapt and evolve in whatever directions are necessary for future evolutionary success.

18

Humanity V. Bacteria

A major reason why most evolutionary biologists think that evolution is not progressive is bacteria. Bacteria seem to be at least as successful as humanity in purely evolutionary terms. They have survived on this planet for over 3,500 million years, they are still surviving, and they seem certain of continued survival, at least in the easily foreseeable future. But against the criteria that humans use to judge each other, bacteria are not very impressive. They are not smart, good looking, rich, or good at basketball. And they are not conscious of either their external environment or of themselves. But the importance to us of these criteria does not justify their use to judge the evolutionary success of all organisms. The ultimate criterion for judging evolutionary success must be survival, and on this, bacteria are doing at least as well as humanity[1].

In fact, many more bacteria are alive on earth now than humans, with about 10 trillion inhabiting a single spoonful of high quality soil. A strong case can be made out that bacteria always have been the dominant form of life on this planet, in terms of both numbers and total mass. Stephen J Gould suggests that we are living in the age of bacteria, and that since life began on earth, it has always been the age of bacteria[2].

But will it always be the age of bacteria in the future? Will bacteria in their current form be flourishing on this planet and elsewhere in the universe in a million years? In a billion years? Will they be the dominant form of life in the solar system, in the galaxy, or in the universe? Or will they be overtaken by other forms of life that will make them insignificant in the future evolution of life in the universe? Is evolution up to now on this planet just the first metre of a 10-kilometre run? Will the forms of life that are winning the race now be overtaken long before the business end of the race?

The fact that many species of bacteria have survived largely unchanged for over 3,500 million years fails to prove that evolution as a whole is not progressive. As I argued in Chapter 2, evolution is fundamentally progressive if there are sequences of potential improvements in living processes that evolution can exploit. The fact that some living processes have not yet exploited the sequences does not mean that the sequences do not exist. It may be that potentials for improvement exist, but the evolutionary mechanisms that have operated until now have not been creative enough to find ways for bacteria to exploit the potentials. This possibility cannot be dismissed without proving that there are no such sequences of potential improvements.

Nevertheless, bacteria do present a considerable challenge to the case for evolutionary progress. It is impossible to argue convincingly that evolution is fundamentally directional and progressive without satisfactorily explaining the success of bacteria. Progressionists must show that the lack of progress in many species of bacteria is not a result of the absence of potentials for improvement. They must point to sequences of potential improvements that will eventually drive progressive evolution once evolution has discovered how to exploit the potentials. Progressionists must demonstrate that bacteria in their current form will eventually be overtaken by new forms of life that are better able to exploit the potentials.

We are now in a position to answer this challenge. Since Chapter 2, we have seen that the benefits of cooperation provide general potentials for on-going improvement that can drive progressive evolution. Whatever evolutionary challenges they face, living processes will be able to respond more effectively through cooperation. As we have seen, the potential benefits of cooperation can be exploited by the formation of managed organisations. The larger the scale of the cooperative organisation, and the higher its evolvability, the more of the benefits of cooperation it will be able to exploit in its pursuit of evolutionary success.

We have seen what sort of living processes are likely to achieve future evolutionary success, and they are not bacteria. Each individual bacterium is a managed cooperative organisation of molecular processes. But individuals of most species of bacteria do not cooperate with each other, or with other organisms. They are not members of large-scale, managed cooperative organisations. The inability of most bacteria to adapt collectively means that they cannot respond effectively to wider-scale events, and cannot have large-scale impacts on their environment. They cannot team up to manage matter, energy, or other living processes on the scale of a centimetre, let alone on the scale of a city, country, nation, planet, solar system or galaxy.

In the long term, bacteria are likely to participate successfully in future evolution only to the extent that they are incorporated into larger-scale cooperative organisations. A few species have already become members of cooperative organisations. Earlier we looked at an example. The mitochondria within the eukaryote cells that form our bodies are the descendants of ancient bacteria. Mitochondria contribute cooperatively to the effective functioning of our cells, and through this to the success of ourselves and our social organisations. They are critically important members of the teams of cells and of the teams of teams of cells that have built the pyramids, invented agriculture, dammed rivers, built cities, and developed genetic engineering and other technologies.

In turn, the mitochondria benefit by being members of organisations that have exploited the benefits of larger-scale cooperation to achieve evolutionary success. Mitochondria by themselves are not able to adapt successfully on the scale that we and our societies do. But as members of our bodies and our social organisations, they share in the benefits that flow from our greater adaptive capacity. Mitochondria are not very smart. They have no capacity for mental modelling. But as members of organisations managed by mental modelling, they evolve and adapt as if they used mental modelling to do so. The management of the organisation controls their adaptation and evolution. If we and our societies use our superior evolvability to pursue evolutionary objectives successfully, we take our mitochondria along with us. Mitochondria have been to the moon and back, and if human organisation ever colonises the galaxy, so to will mitochondria.

Other bacteria are already participating in small-scale cooperative organisations. For example, plants get their essential nitrogen through the cooperative activities of bacteria. Plants cannot use nitrogen directly from the atmosphere. But *Rhizobium* bacteria that live in bulbous growths on the roots of leguminous plants convert nitrogen from the atmosphere into a useable form. As a further example, some bacteria have formed close cooperative relationships with cattle and other grazing animals. These animals could not digest grass without the activities of the bacteria in their stomachs[3].

As we have seen, the next great step forward in the evolution of life on this planet will be the formation of a highly evolvable and cooperative planetary organisation. The bacteria that will be successful in evolutionary terms will be those that become part of this planetary organisation, and that continue their membership as the organisation expands over even wider scales of space and time. Mitochondria are likely to have a role in this organisation while ever it includes humans. But the descendants of other bacteria are also likely to be part of the planetary organisation, at least for a considerable period into the future. This is because bacteria will be critically important to the effective functioning of the planetary organisation, and are likely to be managed by the organisation to enhance their contribution to it. Humanity is currently highly dependent for its survival on the activities of bacteria. We depend for our food, water and oxygen on the effective functioning of the living systems of the planet, and bacteria have a critical role in the operation of these systems. By decomposing dead organic matter, they return to useable form the elements that are essential to life—oxygen, nitrogen, phosphorous, sulphur and nitrogen[4]. Increasingly these bacteria will be managed by humans to optimise their contribution to the success of our societies.

It is not only bacteria that will be incorporated into the planetary organisation in this way. All living processes on the planet that can contribute to the success and expansion of the planetary organisation are likely to be managed as part of the organisation. Of course, we humans already attempt to manage living organisms and living systems when this contributes to achieving our objectives. A typical farm is an organisation of many types of organisms that are managed and controlled by humans to produce food and other resources for our benefit. Farmers use fertilisers and other means to promote the growth of plants that are useful to humans, and eradicate those that are not. They selectively breed animals and crops that are better at satisfying human needs. And farmers take advantage of the ability of bacteria to decompose plant and animal matter to fertilise the soil, and organise their farming practices to promote these bacterial activities where it is advantageous.

Farmers also promote cooperative interactions between organisms where these enhance productivity. For example, dogs are bred and trained to round up sheep, bees and plants are managed in ways that promote fertilisation of plants by bees, and the pattern of cropping is designed to take advantage of the beneficial effects of some crops on others. Management of living systems by humans is not limited to the running of farms. Irrigation projects and pest management schemes are larger-scale attempts to manage living systems for our ends. And on a smaller scale we attempt to manage the bacteria in our bodies and in our local environment to prevent disease.

In many cases, our attempts to manage other living processes have had disastrous consequences. In part this has been because our economic and social systems have failed to ensure that individuals and corporations capture the harmful environmental effects on others of their actions. In part it is because we do not have the knowledge or skills to manage complex living systems successfully. As we improve our systems of governance and accumulate greater knowledge and skills, we should overcome these difficulties. We will be able to manage living processes as part of the planetary organisation so that they contribute to the successful pursuit of evolutionary objectives.

We will redesign the living systems on which we depend, making them more efficient and effective for our purposes. As in farm management, we will manage organisms by changing their genes as well as by controlling their reproduction, environment, and access to food and other resources. Genetic engineering will be used to produce new organisms that function more efficiently and are better able to contribute to the objectives of the planetary organisation. We will use our capacity for mental modelling to continually revise the genetic arrangements of organisms, producing improvements wherever possible. The superior evolvability of humans will be used to exploit potentials for improvement in bacteria and other organisms that the genetic evolutionary mechanism has failed to exploit because of its far more limited evolvability. Organisms that are members of the planetary organisation will be managed so that they adapt and evolve in pursuit of the evolutionary success of the organisation. They will evolve and adapt as if they had the superior evolvability of the organisation as a whole[5].

Bacteria and other organisms will be engineered so that they are better at cooperating together and are managed more easily by the planetary organisation. Where these organisational organisms face damaging competition from free-living organisms that contribute less to the planetary organisation, the organisation will promote the cooperators at the expense of the free-livers. The planetary organisation will determine which living processes continue in existence and flourish, and which will not. Increasingly, free-living bacteria will be out-competed by managed bacteria that benefit from cooperation, or will be suppressed by the planetary organisation itself[6]. It is also likely that human technology will eventually engineer machines that are better than some organisms at undertaking particular functions for the planetary organisation. This might involve the production of teams of microscopic machines that are able to reproduce themselves[7]. These machines might out-compete or directly suppress free-living, unmanaged organisms.

In all these ways, the emergence of a planetary organisation will fundamentally change the living processes on this planet. As the planetary organisation evolves, and as it gets better at managing the living and non-living resources of the planet in its pursuit of evolutionary objectives, there will be vast changes in the types of living processes that inhabit the planet, and in the way they are organised. The evolution of life on earth has often been represented as a growing, branching bush. But as the planetary organisation emerges, this metaphor will no longer be appropriate. Increasingly, the diversity of organisms that has been produced by the branching process will be managed and unified into a single, cooperative organisation of living processes. The branches and twigs of the bush will no longer develop with relative independence. The bush and its growth will be organised and managed into a coherent whole that serves evolutionary objectives.

It is worth emphasising here that this unification of the living processes of the planet into a single organisation will not impose uniformity on them. It will not mean uniformity any more than it did for the groups of cells that were managed to produce multicellular organisms. To the contrary, the formation of managed organisations paves the way for a massive increase in diversity by allowing specialisation and a complex division of labour to emerge. This is what has occurred within cells, within multicellular organisms, and within societies of organisms. In the same way, the planetary organisation will produce unity within difference. The formation of a planetary organisation will facilitate a huge increase in the variety and diversity of living processes, including in human behaviour, and will unify this diversity into a coherent whole.

As the planetary organisation expands into the solar system and the galaxy, it will take with it those living processes that will contribute to its future success. It will leave behind those that will not. Whether or not an organism is retained as part of the expanding organisation will have a critical impact on its future evolutionary significance. Any bacteria and other organisms that are not part of the expanding cooperative organisation will be left to live out the rest of their existence on earth. Cosmologists predict that the sun will eventually expand into a red giant, engulfing most of the planets[8]. When the solar system ends, the organisms that are left behind will also perish, if they survive that long.

In contrast, organisms that remain part of the expanding cooperative organisation will share in any future evolutionary success achieved by the organisation as a whole. They will benefit from the improved evolvability of the expanded organisation, and from its improving ability to manage matter, energy, and life over wider and wider scales of space and time. When the organisation moves from a particular solar system that is soon to collapse, it will take the organisms with it. When it harnesses the energy of a number of stars to enable it to remodel an area of the galaxy to produce a better environment for its purposes, its member organisms will also benefit.

Gould was wrong when he stated that it will always be the age of bacteria. To know what sort of living processes will be successful in future evolution, it is essential to know where evolution is headed. You must know if there are potential improvements in existing living processes that will drive progressive evolution, and, if so, what these improvements are. You must know what the evolutionary game is, and what its winners will look like. We have seen that the evolutionary future will belong to cooperative organisations of living processes that continue to increase in scale and evolvability. If humans change in the right direction both as individuals and socially, we may play a significant role in this future evolution. But we can say with certainty that free-living, unorganised, unmanaged bacteria will not. Non-cooperative, unorganised bacteria have done well in terms of survival up to now. But this is only because it has taken evolution this long to produce an organism that is capable of managing large-scale cooperative organisations of bacteria and other organisms. The future belongs to cooperative organisations, and bacteria will share in this future only to the extent that they participate in and contribute to these organisations[9]. The same applies to us.

19

The Evolutionary Warrior

What contribution should I make to the progressive evolution of humanity? Now that I am aware of the direction of evolution, should I use this knowledge to do what I can to ensure humanity achieves future evolutionary success? Should I promote the formation of a cooperative and highly evolvable planetary organisation, and encourage the development in myself and others of a psychological capacity for evolutionary self-management?

Or should I choose not to direct any of my time and energy to pursuing evolutionary objectives? Since I will not be affected personally by what happens in our distant evolutionary future, should I ignore it? Should I continue to serve the same motivations and goals as I have previously, goals that I now know have been established in me by shortsighted and limited evolutionary mechanisms?

Questions like this will face all individual organisms who become aware of the nature of the evolutionary processes that have formed them and that will determine the evolutionary future of their species.

The way each of us deals with these issues will impact on the ability of humanity to play a significant role in the future evolution of life in the universe. It also has the potential to change how each of us sees ourself, what we do with our life, and the meaning and purpose we see in our individual existence. Evolutionary consciousness has the potential to radically change the experience of being a human being.

Evolutionary awareness impacts on the individual in two main ways.

First, it loosens our attachment to our existing motivations, beliefs, values and objectives. It calls into question the appropriateness of our personal characteristics, and the behaviours they motivate. Second, evolutionary awareness can provide us with new values and objectives, and a new direction to our life. It points to how we can live a life that contributes to the successful evolution of living processes in the universe, a life that is therefore consistent with the forces that are responsible for our existence.

Evolutionary awareness calls into question our existing motivations, beliefs and objectives by showing us what formed them. They are evolution's best attempt until now to get us to behave in ways that will produce evolutionary success. Our most fundamental characteristics have been organised and tailored by evolution for evolutionary ends. Largely, we are the puppets of the evolutionary processes that have produced us. If the evolutionary needs of humanity had been different, the genetic and cultural evolutionary mechanisms would have fitted us out with different motivations, beliefs and values.

This awareness makes it harder for us to continue to take our fundamental characteristics as things that are fixed, given, and beyond conscious choice. Once we see that they have no absolute validity or certainty, we are more likely to see them as things that we can reflect upon and change consciously.

But our attachment to our existing motivations, beliefs and values is shaken more strongly once we begin to see that they fail to do the job they are designed for. Full evolutionary awareness enables us to see the direction of evolution, and what we have to do to achieve future evolutionary success. We can use our mental models of future evolution to test whether we will be successful if we continue to be organised by our existing motivations and needs. Not surprisingly, it is obvious that we will not. The evolutionary mechanisms that have produced us and our behaviour have limited foresight. They have no knowledge of what is needed for future evolutionary success, and have failed to fit us out with the motivations and values needed to achieve it. Evolutionary awareness enables us to see that the motivations and values installed in us by genetic and cultural evolution will fail to produce evolutionary success in the future.

We have been designed for evolutionary success, but poorly. Our mental modelling enables us to see that our current behaviours and social organisation will not deliver future success. We cannot help but ask ourselves some fundamental questions: as far as we can, should we manage and modify our existing motivations, beliefs and values so that they will support the behaviour needed to achieve future evolutionary success? Or should we just continue to serve our pre-existing motivations, knowing that they fail to so the job they are designed for? Can we continue behaviours that we know fail to serve their ultimate objectives? Once we have the ability to evolve through evolutionary mechanisms that have much greater foresight and intelligence than previous mechanisms, can we refuse to use them?

But it is not easy to escape the control exercised over us by our existing motivations, beliefs and values. It is one thing to see mentally that they fail to serve their ultimate objectives. It is another thing entirely to change them. This is because any decision we make about changing our behaviour will be influenced strongly by our existing motivations, beliefs and values. We will tend to use our existing characteristics to decide what we want to do with our lives. And if we have little evolutionary awareness, the evolutionary needs of far-distant generations will not count for much against our more immediate needs for food, sex, money, power and social status.

But evolutionary awareness can change this. It tends to produce individuals who place more value on the evolutionary success of future generations, and less on the gratification of their own immediate urges and needs. Individuals can see that it is the on-going and evolving population of organisms that are important to evolution, not any particular individual. Individuals are significant only to the extent that they contribute to the success of the on-going evolutionary process.

Evolution ensures that individuals that are not self-evolving are temporary. Individuals are evolutionary experiments. There is a chance they will produce improvements that will be passed to the next generation and perpetuated by the species. But if new possibilities are to be tried out and new experiments made, new individuals must be produced. And in a world of limited resources, this means that older individuals must die to make way for the new ones. Older individuals must cease to exist, but the population goes on, made up of new individuals, new experiments.

The evolutionary relevance of individuals is determined by what they can contribute to the evolution of the species. The means by which this contribution is made differs for different evolutionary mechanisms. If the population evolves genetically, individuals can contribute to future generations through their reproduction. If it evolves culturally, individuals can contribute to the on-going process by passing new ideas, discoveries, and other adaptive information to future generations. But from the evolutionary perspective, individuals are irrelevant and meaningless if they contribute nothing to the on-going evolutionary process that produced them, and that will outlive them. They might as well never have lived.

Evolutionary awareness shows us that it is an illusion to see ourselves or other individuals as distinct and separate entities. Individuals are inextricably part of an on-going evolutionary process. They can have no existence without that process. They are born out of it, and can have on-going effects only through what they return to it. In our mental processes, we can separate individuals out from the on-going process and consider them as independent entities. But in reality they are never separate. Without an on-going population that reproduces and evolves through time, there can be no individuals. The population lives in and through the individuals that are its members at any particular time, but it lives on beyond them. Once we can mentally model the processes of life over time scales that are long enough, we cannot help but see that individual organisms such as ourselves are always parts of larger evolving processes.

Mentally we can abstract individuals out of this on-going process and consider them as separate and independent beings. If we do this and then ask what purpose or meaning there can be in their temporary lives, we must conclude that there can be none. From the point of view of an isolated individual, the inevitability of death makes anything he does during his life irrelevant and meaningless. If there is no afterlife, whatever he does during his short life can make no difference to himself in the long run. He still ends up dead. It is impossible to find meaning in a temporary and isolated life that is not part of some on-going process.

This is the essence of the type of thinking that has produced the collapse of meaning and purpose in the lives of many people this century. But this thinking is wrong because its starting point is wrong. It begins with individuals who are artificially abstracted from the on-going evolutionary process. Evolutionary awareness enables us to avoid this error. It correctly sees individuals as part of an on-going and evolving process. This process continually produces new individuals and fits them out with the adaptive knowledge accumulated by the process over the billions of years it has evolved. The lives of countless organisms who have come before us have contributed to this accumulation. During our own lives we have the chance to consciously add to this knowledge by discovering and implementing adaptive improvements, and contributing them to the on-going process. The process goes on after our death, producing new individuals who are equipped with any improvements we have contributed. In this broader context individuals are the method used by the on-going process to perpetuate itself and to discover more about how to achieve evolutionary success. Abstracted from this process, our lives make as much sense as would the lives of our cells if their relationship with our bodies is ignored.

We can live our lives as if we are separate from the on-going evolutionary process, and make no conscious contribution to it. But individuals with evolutionary awareness will not find meaning or purpose in such a life. They will find no meaning in a life spent vigorously and energetically seeking the satisfaction of their pre-existing material and emotional urges. A life dedicated to the pursuit of money, sex, power, and social status within our current social structures is a wasted life from the perspective of evolutionary awareness. Meaning can come only from taking whatever opportunities we have to contribute to the on-going evolutionary processes that will outlive us.

But if we are to escape control by the motivations, needs, beliefs and values established in us by shortsighted evolutionary processes, we need to discover how to modify and manage these internal predispositions. We must be able to consciously manage our predispositions and our external environment so that we can find satisfaction and motivation in whatever we need to do to contribute to future evolutionary success. We must develop the psychological capacity to transcend our biological and cultural predispositions so that we can consciously adapt our behaviour in whatever way we see is necessary. Our ultimate objectives will be set by our mental modelling of future evolutionary possibilities, and we must be able to do whatever is suggested by these models. Our adaptive repertoire must not be restricted by the motivations, needs, beliefs and values established by past evolution.

Evolutionary awareness shows us that this is a challenge faced by all organisms that develop the capacity to model their evolutionary future. Any organism that develops evolutionary awareness will see that they must transcend their biological and cultural past if they are to be able to do what is necessary for future evolutionary success. They know that the organisms who will contribute most to the evolution of life in the universe are those that meet this challenge successfully. Evolutionary awareness also shows us that our struggle to develop the psychological skills needed to transcend our social and biological past is part of an evolutionary event of great significance on this planet. And it makes us aware that our growing evolutionary awareness is itself an important part of the unfolding of this evolutionary event.

As we have seen, humans have already begun to develop the psychological skills that will enable us to manage our pre-existing adaptive processes consciously. Once organisms evolve a capacity for mental modelling, it is inevitable that they will begin to learn these skills. Because mental modelling has superior foresight, it will often discover better adaptations than those established by the pre-existing adaptive processes. As the modelling capacity improves, it increasingly discovers circumstances in which existing motivations and emotional urges produce behaviour that are against the longer-term interests of the organism. To take advantage of this superior foresight, organisms begin to learn how to ensure that their behaviour is guided by their mental models, not by their pre-existing adaptive processes.

So we begin to learn to manage our sexual urges, anger and other emotional impulses when we see that it is in our longer-term interests to do so. For example, many of us learn to forgo the immediate gratification of some of our needs in order to achieve longer-term career goals. We learn to organise our thoughts, motivations and our environment to find satisfaction in study and other behaviours that serve these career goals. The logical extension of this capacity would be the ability to manage our pre-existing adaptive processes to make them consistent with our interests over all time scales, including over evolutionary scales.

As we have seen, we must turn our mental modelling inwards if we are to develop fully the psychological capacity to consciously modify and manage our pre-existing physical, emotional and mental adaptive processes. Our thoughts, beliefs, motivations and emotional states, as well as the methods we use to influence them, must become objects of consciousness. This would provide us with the self-knowledge and self-awareness needed to develop the capacity to manage our internal adaptive processes. It would enable us to develop a new "I" that is not limited or controlled in what it can do by our biological and cultural past. The new "I" would act upon our internal processes, rather than being acted upon by them. Our reactions to external circumstances would no longer be determined automatically by our pre-existing mental and emotional predispositions. Instead our new "I" would be able to decide to intervene before these predispositions determined our reactions. Our new "I" would then be able to consciously organise a different response that is more appropriate to our evolutionary objectives. It would see what we must do to contribute to future evolutionary success and would produce this behaviour by consciously managing our internal processes, and intervening in them. The result would be an organism that is able to recreate and reinvent itself through conscious choice as often as is necessary to meet evolutionary demands.

The development of this new psychological software would free individuals from their biological and cultural past. It would produce individuals who are self-evolving and who therefore have a completely new evolutionary status. We saw earlier that individuals who are not self-evolving must be temporary if evolution is to proceed successfully. Individuals with restricted adaptability and evolvability would stall evolution if they live forever. For this reason, organisms who are not self-evolving but whose overriding objective is to contribute to the successful evolution of life would not attempt to achieve immortality. To do so would contradict their fundamental purpose.

But a being that is truly self-evolving would not impede evolution if it lived forever. If an individual were truly capable of adapting in whatever ways are demanded by future evolutionary needs, it would not have to die for evolution to proceed. Evolution could try out new possibilities without having to replace the individual. New possibilities could be fully explored within a self-evolving individual. The internal adaptive processes of self-evolving beings have the capacity to adapt freely in any direction. With the emergence of self-evolving individuals, the evolutionary process would be internalised within the individual.

The acquisition of evolutionary awareness would also show each of us that if humanity is to achieve future evolutionary success, we must not only transform ourselves, we must also transform our societies. Our individual efforts to contribute to the successful evolution of life will be futile without radical changes to our social organisation. It is only through the formation of cooperative organisations of larger scale and greater evolvability that humanity can participate successfully in the future evolution of life in the universe. We must continue the progressive evolution on this planet that has seen molecular processes organised into cells, cells organised into multicellular organisms, and organisms organised into societies.

We have seen that a significant step forward would be the formation of a cooperative planetary society whose ultimate objective is to pursue future evolutionary success. The society would manage the matter, energy and living resources of the planet in whatever ways would advance its evolutionary interests. The management of the planetary society would support and reward behaviours that assisted the pursuit of evolutionary objectives. As a result, cooperation that promoted evolutionary success would pay. The members of the society would capture the evolutionary effects on others of their actions. Individuals would find it advantageous to do whatever was necessary to enhance the evolutionary success of the planetary society. And individuals would be provided with feedback about what actions would contribute most to the evolutionary success of the society. Global governance would produce a cooperative planetary society in the same way that the governance of molecular processes produced the cooperative organisations that became cells, the governance of cells produced the cooperative groups of cells that became multicellular organisms, and the governance of organisms produced societies of organisms.

But our current economic systems and forms of governance are not capable of organising a planetary society that can fully exploit the benefits of cooperation in the interests of its members. Our governance and our economic systems do not always ensure that individuals capture the effects of their actions on others. And the evolvability of these systems can be substantially improved. Evolutionary awareness tells us that we must transform our current economic systems and forms of governance if we are to achieve future evolutionary success, and identifies the types of changes that need to be made. It shows us how to build cooperative and evolvable organisations out of self-interested components. As we have seen, an important step is to enhance the adaptability and effectiveness of our systems of governance by implementing a vertical market system and the supra-individual change-and-test processes that go with such a market.

The result would be a highly evolvable planetary society that efficiently and adaptively manages the matter, energy and living processes of the planet to serve the objectives of the members of the society. But these changes will not produce a society that pursues future evolutionary success until most of its members also embrace this objective. If its members continue to be motivated by pre-existing biological and cultural needs, the society and all its resources will be harnessed to serve those needs. The society will serve needs that are inconsistent with the continued evolutionary success of life on earth. If life on this planet does not progress beyond this point, it will be a failed evolutionary experiment.

But when the majority of members of the society embrace evolutionary objectives, so to will the society. The society will begin to manage the matter, energy and living processes of the planet to achieve evolutionary objectives. In its pursuit of future evolutionary success, the planetary society will begin to develop an ability to adapt and act as a coherent whole. The planetary society will develop the ability to make plans, develop projects, establish objectives, and deal with any alien societies it encounters. A society of evolutionary warriors will eventually self-actualise as an evolutionary warrior in its own right.

Against this background, we can now summarise what individuals at the beginning of the 21st century should do to contribute to the future evolutionary success of life on this planet. Those of us who acquire evolutionary awareness will have three general projects: first, we will work on ourselves to improve our adaptability and evolvability. Our objective will be to develop the self-knowledge and psychological skills needed to transcend our biological and cultural past. We will develop a psychology that is no longer controlled by the needs and motivations we inherit biologically and culturally. This psychology will enable us to be self-evolving. It will equip us with the capacity to find motivation and satisfaction in whatever we need to do to contribute to evolutionary objectives.

Second, we will promote in others a deeper understanding of the evolutionary process, and will encourage and assist them to develop the psychological capacity to use this understanding to guide and manage their own behaviour. Humanity cannot make significant progress in evolutionary terms until the majority of us embrace evolutionary objectives. Future evolutionary success can only be achieved collectively and cooperatively. And if human society as a whole is to be an evolutionary warrior that transcends its biological and cultural past, its members must first become evolutionary warriors. The evolvability of human society will depend on the evolvability and adaptability of its members. Those of us that become aware of this will promote evolutionary awareness and self-management in others. One of the best ways we can do this is to develop these capacities to the fullest extent in ourselves.

Finally, we will support the formation of a unified and self-actualised planetary society. We will do what we can to develop a society that uses its understanding of the direction of evolution to guide its own evolution. Critical steps along the way to achieving such a society will be the establishment of a system of global governance, and the implementation of new forms of governance and economic systems that are more evolvable and better at organising cooperation. The society will develop and utilise the potential of all its members in order to fulfil its own potential.

It would over-dramatise the situation to say that the future evolutionary success of humanity depends on how well we pursue these projects. We can hope that others who come after us will have the opportunity to repair any damage we cause, and make up for any short fall in what we achieve. But how successful we are at advancing these projects will determine the significance of our lives.

Notes and References

Chapter 1 Introduction

1. A clear discussion of the distinction between human and evolutionary progress can be found in Ayala, F. J. (1988) Can "Progress" be Defined as a Biological Concept? In *Evolutionary Progress*. (Nitecki, M. H. ed.) pp 75-96. Chicago: University of Chicago Press.
2. The great majority of the leading evolutionary theorists who attended a major international conference on evolutionary progress in 1988 opposed the view that evolution is progressive and that humans are at the leading edge of evolution on this planet. The key papers delivered at the conference are in Nitecki, M. H. ed. (1988) *Evolutionary Progress*. Chicago: University of Chicago Press.
3. For example, see: Blitz, D. (1992) *Emergent Evolution: Qualitative Novelty and the Levels of Reality*. Dordrecht: Kluwer Academic Publishers; and Ruse, M. (1996) *Monad to Man*. Cambridge, MA: Harvard University Press.
4. See Gould, S. J. (1996) *Full House: the Spread of Excellence from Plato to Darwin*. New York: Harmony Books; and Maynard Smith, J. (1988) Levels of Selection. In *Evolutionary Progress*. (Nitecki, M. H. ed.) pp 219-236. Chicago: University of Chicago Press.

5. For a history of evolutionary thought, see Bowler, P. J. (1984) *Evolution: The History of an Idea.* Berkeley: University of California Press.

6. These ideas were first developed in detail in Stewart, J. E. (1995) Metaevolution. *Journal of Social and Evolutionary Systems* **18**: 113-47; Stewart, J. E. (1997) Evolutionary transitions and artificial life. *Artificial Life* **3**: 101-120; and Stewart, J. E. (1997) Evolutionary Progress. *Journal of Social and Evolutionary Systems* **20**: 335-362.

7. For example, see Dawkins, R. (1976) *The Selfish Gene.* New York: Oxford University Press; and Williams, G. C. (1966) *Adaptation and Natural Selection.* Princeton: Princeton University Press.

8. Our discussion in Chapter 9 will build on the ideas outlined in Stewart, J. E. (1997) The Evolution of Genetic Cognition. *Journal of Social and Evolutionary Systems* **20**: 53-73; and in Moxon, R. E. and Thaler, D. S. (1997) The Tinkerer's Evolving Tool-box. *Nature* **387**: 659-662.

9. See Stewart, J. E. (1993) The Maintenance of Sex. *Evolutionary Theory.* **10**: 195-202; and Stewart: The Evolution of Genetic Cognition. *op. cit.*

10. Our discussion of the evolution of internal adaptive mechanisms in Chapter 10 will build on the ideas developed in Dennett, D. C. (1995) *Darwin's Dangerous Idea.* New York: Simon and Schuster; Stewart: Metaevolution. *op. cit*; and Stewart: Evolutionary Progress. *op. cit.*

11. The significance of being able to first try out innovative behaviour mentally rather than in practice is emphasised by Popper, K. R. (1972) *Objective knowledge—an evolutionary approach.* Oxford: Clarendon.

12. See, for example, Klein, R. G. (1989) *The Human Career: Human Biological and Cultural Origins.* Chicago: University of Chicago Press; Boehm, C. (1993) Egalitarian Behaviour and Reverse Dominance Hierarchy. *Current Anthropology.* **34**: 227-54; and Knauft, B. M. (1991) Violence and Sociality in Human Evolution. *Cur-*

rent Anthropology. **32**: 391-428.

13. The discussion in Chapters 11 and 12 will build on the ideas first developed in Stewart: Metaevolution. *op. cit*; and Stewart: Evolutionary Progress. *op. cit.*
14. A well-written broad sketch of the evolution of human society can be found in Chapter 14 (pages 265-292) of Diamond, J. (1997) *Guns, Germs and Steel.* London: Jonathon Cape.

Chapter 2 The Causes of Progress

1. For a readable, comprehensive and insightful account of the progressive evolution of communication and related technologies see Levinson, P. (1997) *A Natural History and Future of the Information Revolution.* Routledge: New York.
2. See, for example, Chapter 15 of Gould, S. J. (1996) *Full House: the Spread of Excellence from Plato to Darwin*. New York: Harmony Books.
3. Loftas, T. Ed. (1995) *Dimensions of need: an atlas of food and agriculture.* Rome, Italy: Food and Agriculture Organization of the United Nations.
4. For a detailed argument in favour of this position see Durham, W. H. (1991) *Coevolution: Genes, Culture and Human Diversity.* Stanford, CA: Stanford University Press.
5. Dawkins, R. (1997) Human chauvinism. *Evolution* **51**: 1015-1020.
6. Dawkins, R., and J. R. Krebs. (1979) Arms races between and within species. *Proceedings of the Royal Society, London. B Biol. Sci.* **205**: 489-511.
7. For example, see Carter, G. S. (1967) *Structure and Habit in Vertebrate Evolution*. London: Sidgwick and Jackson.
8. For example, see Ayala, F. J. (1997) Ascent by natural selection. *Science* **275**: 495-6; and Gould: *Full House: the Spread of Excellence from Plato to Darwin*. *op. cit.*
9. See Chapter 7 of Stebbins, G. L. (1971) *Processes of Organic Evolution*. New Jersey: Prentice-Hall.

Chapter 3 Why Cooperate?

1. For a comprehensive picture of the functioning of multicellular organisms from an integrated systems perspective see Miller, J. G. (1978) *Living Systems*. McGraw-Hill.
2. For a very clear account of the structure and functioning of cells see Rensberger, B. (1996) *Life itself: exploring the realm of the living cell.* Oxford: Oxford University Press.
3. See Chapter 2 of Ridley, M. (1996) *The Origins of Virtue*. London: Viking.
4. A comprehensive picture of differentiation, specialisation and integration at all levels of living processes is provided by Miller: *Living Systems. op. cit.*
5. See Corning, P. (1998) The Cooperative Gene: On the Role of Synergy in Evolution. *Evolutionary Theory* **11**: 183-207.
6. Many examples of advantageous cooperation between organisms are given by Dugatkin, L. (1999) *Cheating Monkeys and Citizen Bees: The Nature of Cooperation in Animals and Humans.* New York: Free Press.
7. See the classic paper Hardin, G. (1968) The tragedy of the commons. *Science* **162**: 1243-8.

Chapter 4 Barriers to Cooperation

1. Loftas, T. Ed. (1995) *Dimensions of need: an atlas of food and agriculture.* Rome, Italy: Food and Agriculture Organization of the United Nations.
2. Williams, G. C. (1966) *Adaptation and Natural Selection.* Princeton: Princeton University Press.
3. Dawkins, R. (1976) *The Selfish Gene.* New York: Oxford University Press.
4. Williams: *Adaptation and natural selection. op. cit.*
5. Maynard Smith, J., and Szathmary, E. (1995) *The Major Transitions in Evolution.* Oxford: W. H. Freeman; and Stewart, J. E. (1995) Metaevolution. *Journal of Social and Evolutionary Sys-*

tems **18**: 113-147.

6. Olson, M. (1965) *The Logic of Collective Action.* Cambridge, Mass: Harvard University Press.
7. Buss, L. W. (1987) *The Evolution of Individuality.* Princeton: Princeton University Press.
8. Margulis, L. (1981) *Symbiosis in cell evolution.* San Francisco: W. H. Freeman.
9. Frank, S. A. (1997) Models of symbiosis. *American Naturalist* **150**: S80-S99.
10. Eberhard, W. G. (1980) Evolutionary consequences of intracellular organelle competition. *Quarterly Review of Biology* **55**: 231-249.
11. Hanson, M. R. (1991) Plant mitochondrial mutations and male sterility. *Annual Review of Genetics* **25**: 461-486.
12. Farmer, J. D., Kauffman, S. A., and N. H. Packard (1986) Autocatalytic replication of polymers. *Physica D*, **22**: 50-67.
13. Kauffman, S. A. (1993) *The Origins of Order: Self-organisation and selection in evolution.* New York: Oxford University Press; and Kauffman, S. A. (1995) *At home in the universe: The search for laws of self-organisation and complexity.* New York: Oxford University Press.
14. Bagley, R. J. and Farmer, J. D., (1991) Spontaneous Emergence of a Metabolism. In: *Artificial Life II*. (Langton, C. et al, eds.) New York: Addison and Wesley; and Maynard Smith, J. (1979) Hypercycles and the origin of life. *Nature, Lond.* **280**: 445-446.

Chapter 5 Organising Cooperation

1. For the first attempts to build a comprehensive theory of the evolution of cooperation across all these levels, see Maynard Smith, J., and E. Szathmary (1995) *The Major Transitions in Evolution.* Oxford: W. H. Freeman; Stewart, J. E. (1995) Metaevolution. *Journal of Social and Evolutionary Systems* **18**: 113-147; and Stewart, J. E. (1997) Evolutionary Progress. *Journal of Social and Evolu-*

tionary Systems **20**: 335-362.

2. Hamilton, W. (1964) The Genetical Evolution of Social Behaviour. *Journal of Theoretical Biology.* **7:** 1-52.
3. For example, see Chapter 9 of Brown, J. L. (1975) *The Evolution of Behaviour.* New York: W. W. Norton.
4. Brown, J. L. (1970) Cooperative breeding and altruistic behaviour in the Mexican jay. *Animal Behaviour* **18**: 366-378.
5. Heinsohn, R. G. (1992) Cooperative enhancement of reproductive success in white-winged choughs. *Evolutionary Ecology* **6**: 97-114.
6. See Corning, P. (1998) The Cooperative Gene: On the Role of Synergy in Evolution. *Evolutionary Theory* **11**: 183-207.
7. For example, see Wilson, E. O. (1975) *Sociobiology: The New Synthesis.* Cambridge, MA: Harvard University Press.
8. Trivers, R. (1972) The Evolution of Reciprocal Altruism. *Quarterly Review of Biology* **46**: 35-57; and Axelrod, R. and D. Dion (1989) The further evolution of cooperation. *Science.* **232**: 1385-1390.
9. Wilkinson, G. S. (1984) Reciprocal food sharing in the vampire bat. *Nature* **308**: 181-184.
10. Axelrod, R. (1985) *The Evolution of Cooperation.* New York: Basic Books.
11. Martinez-Coll, J. C. and Hirshleifer, J. (1991) The limits of reciprocity. *Rationality and Society* **3**: 35-64.
12. Dugatkin, L. (1999) *Cheating Monkeys and Citizen Bees: The Nature of Cooperation in Animals and Humans.* New York: Free Press.
13. See Dunbar, R. (1996) *Grooming, Gossip and the Evolution of Language.* London: Faber and Faber.
14. Stewart, J. E. (1997) Evolutionary transitions and artificial life. *Artificial Life* **3**: 101-120.
15. Stewart: Metaevolution. *op. cit.*; and Stewart: Evolutionary transitions and artificial life. *op. cit.*

16. Stewart: Metaevolution. *op. cit.*; and Stewart: Evolutionary transitions and artificial life. *op. cit.*

17. This impediment to the evolution of cooperation is termed the 'second order problem' by Axelrod, R. (1986) The evolution of norms. *American Political Science Review* **80**: 1095-1111.

18. Stewart: Metaevolution. *op. cit.*

19. Stewart: Metaevolution. *op. cit.*; Stewart: Evolutionary transitions and artificial life. *op.cit.*; and McGuire, M. C. and M. Olson (1996) The Economics of Autocracy and Majority rule: The Invisible Hand and the Use of Force. *Journal of Economic Literature* **34**: 72-96.

20. Hodgson, G. (1988) *Economics and Institutions.* Cambridge: Polity Press.

21. Frank, S. A. (1997) Models of symbiosis *American Naturalist* **150**: S80-S99.

22. Eberhard, W. G. (1980) Evolutionary consequences of intracellular organelle competition. *Quarterly Review of Biology* **55**: 231-249.

23. Eberhard: Evolutionary consequences of intracellular organelle competition. *op. cit.*

24. Blackstone, N. W. (1995) A units-of-evolution perspective on the endosymbiont theory of the origin of the mitochondrion. *Evolution* **49**: 785-796.

25. Stewart: Metaevolution. *op. cit.*

26. Wilkinson: Reciprocal food sharing in the vampire bat. *op. cit.*

27. Brown, J. L. (1987) *Helping and communal breeding in birds: ecology and evolution.* Princeton: Princeton University Press.

28. Nowak, M. A., May, R. M. and K. Sigmund (1995) The Arithmetics of Mutual Help. *Scientific American* June 1995: 50-55.

Chapter 6 The Evolution of Management

1. Stewart, J. E. (1995) Metaevolution. *Journal of Social and Evolutionary Systems* **18**: 113-147; and Stewart, J. E. (1997) Evolutionary transitions and artificial life. *Artificial Life* **3**: 101-120.

2. Stewart: Metaevolution. *op. cit.*; and Stewart: Evolutionary transitions and artificial life. *op. cit.*
3. Dyson, F. (1985) *Origins of life.* London: Cambridge University Press.
4. Stewart: Metaevolution. *op. cit*; and Stewart: Evolutionary transitions and artificial life. *op. cit.*
5. Maynard Smith, J. (1979) Hypercycles and the origin of life. *Nature, Lond.* **280**: 445-446; and Bresch, C., Niesert, U., and Harnasch, D. (1979) Hypercycles, parasites and packages. *Journal of Theoretical Biology* **85:** 399-405.
6. Wilson, D. S. and Sober, E. (1989) Reviving the Superorganism. *Journal of Theoretical Biology* **136:** 337-356.
7. Bresch, Niesert, Harnasch: Hypercycles, Hypercycles, parasites and packages. *Op. cit.*
8. Maynard Smith, J. and E. Szathmary (1993) The origin of chromosomes I. Selection for linkage. *Journal of Theoretical Biology.* **164**: 437-446.
9. Ettinger, L. (1986) Meiosis: a selection stage preserving the genome's pattern of organisation. *Evolutionary Theory.* **8:** 17-26.
10. Stewart: Evolutionary transitions and artificial life. *op. cit.*; and Leigh, E. (1977) How does selection reconcile individual advantage with the good of the group? *Proc. Natl. Acad. Sci. USA* **74**: 4542-4546.
11. Stewart: Metaevolution. *op. cit*; and McGuire, M. C. and Olson, M. (1996) The Economics of Autocracy and Majority rule: The Invisible Hand and the Use of Force. *Journal of Economic Literature* **34**: 72-96.
12. For example, see Hamilton, E. (1993) *The Roman Way.* New York: W. W. Norton.
13. Stewart: Evolutionary transitions and artificial life. *op. cit.*
14. Stewart: Metaevolution. *op. cit.*
15. Orwell, G. (1946) *Animal Farm.* New York: Harcourt Brace Jovanovich.

16. Quoted at page 34 in Thurow, L. (1997) *The Future of Capitalism.* St Leonards: Allen and Unwin.

17. See, for example, Mc Murtry, J. (1978) *The Structure of Marx's World-View.* Princeton, New Jersey: Princeton University Press.

18. Stewart: Metaevolution. *op. cit.*; and Stewart, J. E. (1997) Evolutionary Progress. *Journal of Social and Evolutionary Systems* **20**: 335-362.

19. Miller, S. L. (1953) Production of some organic compounds under possible primitive Earth conditions. *J. Am. Chem. Soc.* **77**: 2351-2360; and Fox, S. W. (1988) *The emergence of life: Darwinian evolution from the inside.* New York: Basic Books.

20. Stewart: Evolutionary transitions and artificial life. *op. cit.*

21. Kauffman, S. A. (1993) *The Origins of Order: Self-organisation and selection in evolution.* New York: Oxford University Press.

22. Stewart: Evolutionary transitions and artificial life. *op. cit.*

Chapter 7 Internal Management

1. Buss, L. W. (1987) *The Evolution of Individuality.* Princeton: Princeton University Press.

2. Wilson, E. O. (1975) *Sociobiology: The New Synthesis.* Cambridge, MA: Harvard University Press.

3. Boehm, C. (1993) Egalitarian Behaviour and Reverse Dominance Hierarchy. *Current Anthropology* **34**: 227-54.

4. Stewart, J. E. (1995) Metaevolution. *Journal of Social and Evolutionary Systems* **18**:113-147; and Stewart, J. E. (1997) Evolutionary Progress. *Journal of Social and Evolutionary Systems* **20**: 335-362.

5. Boyd, R. and Richerson, P. (1992) Punishment Allows the Evolution of Cooperation (or anything else) in sizable groups. *Journal of Ethology and Sociobiology* **13**: 171-195.

6. Stewart, J. E. (1997) Evolutionary transitions and artificial life. *Artificial Life* **3**: 101-120.

7. Alexander, R. D. (1987) *The biology of moral systems.* New York:

Aldine de Gruyter.

8. Hamilton, W. (1964) The Genetical Evolution of Social Behaviour. *Journal of Theoretical Biology.***7:** 1-52.
9. Heinsohn, R. G. (1994) Helping is costly to young birds in cooperatively breeding white-winged choughs. *Proc. R. Soc. Lond. B* **256**: 293-298.
10. Stewart: Evolutionary transitions and artificial life. *op. cit.*
11. Holldobler, B. and E. O. Wilson (1990) *The Ants.* Cambridge, MA: Harvard University Press.
12. Heinze, J., Holldobler, B. and C. Peeters (1994) Conflict and cooperation in ant societies. *Naturwissenschaften* **81**: 489-497.
13. Ratnieks, F. L. and P. K. Visscher (1989) Worker policing in the honeybee. *Nature, Lond.* **342**: 796-797.
14. Heinze, Holldobler, and Peeters: Conflict and cooperation in ant societies. *op. cit.*
15. Buss, L. W. (1987) *The Evolution of Individuality.* Princeton: Princeton University Press.
16. Maynard Smith, J., and Szathmary, E. (1995) *The Major Transitions in Evolution.* Oxford: W. H. Freeman.

Chapter 8 Smarter Cooperation

1. Stewart, J. E. (1995) Metaevolution. *Journal of Social and Evolutionary Systems* **18**: 113-147.
2. Stewart: Metaevolution. *op. cit.*; and Stewart, J. E. (1997) Evolutionary Progress. *Journal of Social and Evolutionary Systems* **20**: 335-362.
3. What I label here as a 'change-and-test process' is essentially the same as what is called a 'generate-and-test process' in Dennett, D. C. (1995) *Darwin's Dangerous Idea.* New York: Simon and Schuster.
4. McFarland, D. (1985) *Animal behavior: psychobiology, ethology, and evolution.* Menlo Park, California: Benjamin/Cummings Pub. Co.

5. See Skinner, B. F. (1953) *Science and Human Behaviour.* New York: Knopf.

6. Seeley, T. D. (1995) *The Wisdom of the Hive*. Cambridge, MA: Harvard University Press.

7. For example, see Haveman, R. H. and K. A. Knopf (1966) *The Market System.* New York: John Wiley and Sons.

8. Skinner: *Science and Human Behaviour. op. cit.*

9. For example, see Popper, K. R. (1972) *Objective knowledge - an evolutionary approach.* Oxford: Clarendon; Dennett: *Darwin's Dangerous Idea. op. cit.*; and Stewart: Metaevolution. *op. cit.*

Chapter 9 Smarter Genes

1. Bagley, R. J. and Farmer, J. D., (1991) Spontaneous emergence of a metabolism. In: *Artificial Life II.* (Langton, C. et al, eds.) New York: Addison and Wesley.

2. Kauffman, S. A. (1993) *The Origins of Order: Self-organisation and selection in evolution.* New York: Oxford University Press.

3. *Ibid.*

4. *Ibid.*

5. Stewart, J. E. (1995) Metaevolution. *Journal of Social and Evolutionary Systems* **18**: 113-147.

6. *Ibid.*

7. Wilson, D. S. and Sober, E. (1989) Reviving the Superorganism. *Journal of Theoretical Biology* **136:** 337-356.

8. For example, see Dawkins, R. (1986) *The Blind Watchmaker.* London: Longmans.

9. Williams, G. C. (1966) *Adaptation and natural selection. A critique of some current evolutionary thought.* Princeton University Press: Princeton.

10. Karlin, S. and McGregor, J. (1974) Towards a theory of the evolution of modifier genes. *Theoretical Population Biology* **5**: 59-103; and Lieberman, U. and Feldman, M. W. (1986) Modifiers of mutation rate: a general reduction principle. *Theoretical Population*

Biology **30**: 125-142.

11. Provided, of course, the mutator and the mutated gene are sufficiently closely linked on the same chromosome for the mutator to share in the success of the mutated gene.
12. Leigh, E. G. (1973) The evolution of mutation rates. *Genetics* **73** (Supplement): 1-18; and Kondrashov, A. S. (1988) Deleterious mutations and the evolution of sexual reproduction. *Nature* **336**: 435-440.
13. Ghiselin, M. T. (1974) *The economy of nature and the evolution of sex.* University of California Press: Berkeley, California; Williams, G. C. (1975) *Sex and evolution.* Princeton University Press: Princeton, New Jersey; and Hamilton, W. D. (1980) Sex versus non-sex versus parasite. *Oikos* **35:** 282-290.
14. Jaenike, J. (1978) An hypothesis to account for the maintenance of sex within populations. *Evolutionary Theory* **3:** 191-4; and Hamilton, W. D. (1980) Sex versus non-sex versus parasite. Oikos **35:** 282-290.
15. Hamilton, W. D., Axelrod, R. and Tanese, R. (1990) Sexual reproduction as an adaptation to resist parasites. Proc. Nat. Acad. Sci. USA **87**: 3566-3573.
16. Stewart, J. E. (1993). The maintenance of sex. *Evolutionary Theory.* **10:** 195-202.
17. Gillespie, J. H. and Turelli, M. (1989) Genotype-environment interactions and the maintenance of polygenic variation. *Genetics* **121**: 129-138.
18. Stewart, J. E. (1993) The maintenance of sex. *Evolutionary Theory.* **10:** 195-202.
19. Leigh, E. G. (1970) Natural selection and mutability. *Am. Nat.* **104**: 301-305; and Leigh: The evolution of mutation rates. *op. cit.*
20. Ishii, K., Matsuda, H., Iwasa, Y. and Sasaki, A. (1989) Evolutionary stable mutation rate in a periodically changing environment. *Genetics* **121**: 163-174; and Stewart, J. E. (1997) The Evolution of Genetic Cognition. *Journal of Social and Evolutionary Systems*

20: 53-73.

21. Stewart, J. E. (1997) The Evolution of Genetic Cognition. *Journal of Social and Evolutionary Systems* **20**: 53-73.
22. Of course, a mutator could also survive a period of stability if it produced no mutations during the period, and if it commenced producing mutations only once it was 'switched on' by relevant evironmental change. There is little clear evidence yet that such mechanisms have evolved, but it is early days. See Radman, M. (1999) Enzymes of evolutionary change. *Nature* **401**: 866-9.
23. Stewart: The Evolution of Genetic Cognition. *op. cit.*
24. *Ibid.*
25. Moxon, R. E. and Thaler, D. S. (1997) The Tinkerer's Evolving Tool-box. *Nature* **387**: 659-662; and Brookes, M. (1998) Day of the Mutators. *New Scientist* 14 February 1998: 38-42. For references to other examples, see Pennisi, E. (1998) Molecular evolution—how the genome readies itself for evolution. *Science*. **281**: 1131-1134.
26. Stewart: The Evolution of Genetic Cognition. *op. cit.*
27. Stewart: The Maintenance of Sex. *op. cit*; and Stewart: The Evolution of Genetic Cognition. *op. cit.*
28. Stewart: The Evolution of Genetic Cognition. *op. cit.*
29. *Ibid.*
30. For a fuller discussion of how these factors can influence the rate at which variation is produced and its content see Stewart: The Evolution of Genetic Cognition. *op. cit*; and Mather, K. (1943) Polygenic inheritance and natural selection. *Biological Reviews* **18**: 32-64.
31. Evolution will favour controller genes that are linked closely enough to their effects to capture the benefits they create.
32. Brooks, L. D. (1988) The evolution of recombination rates. In: *The evolution of sex* (Michod, R. E. and Levin, B. eds.) Sunderland Mass.: Sinauer; Lichten, M. and Goldman, A. S. H. (1995) Meiotic recombination hotspots. *Annu. Rev. Genetics.* **29**: 423-444:

and Pennisi: Molecular evolution—how the genome readies itself for evolution. *op. cit.*

33. Wagner, P. W. and Altenberg, L. (1996) Complex adaptations and the evolution of evolvability. *Evolution* **50**: 967-976.
34. Gerhart, J. and M. Kirschner (1997) *Cells, Embryos and Evolution.* New York: Blackwell Science Inc.
35. Wagner and Altenberg: Complex adaptations and the evolution of evolvability. *op. cit.*
36. Dawkins, R. (1989) The evolution of evolvability. In: *Artificial Life.* (Langton, C. ed.) New York: Addison and Wesley; and Gerhart and Kirschner: *Cells, Embryos and Evolution. op. cit.*
37. Dawkins: The evolution of evolvability. *op. cit.*

Chapter 10 Smarter Organisms

1. Stewart, J. E. (1995) Metaevolution. *Journal of Social and Evolutionary Systems* **18**: 113-147.
2. Thorpe, W. H. (1956) *Learning and instinct in animals.* London: Methuen.
3. Ashby, W. R. (1964) *An introduction to cybernetics.* 2nd ed. Chapman and Hall.
4. Ashby, W. R. (1960) *Design for a Brain.* 2nd ed. New York: Wiley.
5. Hardy, R. N. (1982) *Homeostasis.* London: Edward Arnold.
6. Beer, S. (1966) *Decision and control.* New York: John Wiley and Sons.
7. Beer, S. (1972) *Brain of the firm.* London: Allen Lane.
8. *Ibid.*
9. Stewart: Metaevolution. *op. cit*; and Stewart, J. E. (1997) Evolutionary Progress. *Journal of Social and Evolutionary Systems* **20**: 335-362.
10. Thorpe: *Learning and instinct in animals. op. cit.*
11. See Alcock, J. (1993) 5th Edition. *Animal behaviour: an evolutionary approach.* Sunderland, Massachusetts: Sinauer Associates,

Inc.

12. For example, see Ridley, M. (1986) *Animal Behaviour*. London: Blackwell Scientific Publications.
13. Frank, R. H. (1988) *Passions within Reason*. New York: Norton; and Stewart: Evolutionary Progress. *op. cit*.
14. The internal reward systems are the innate teaching mechanisms of Lorenz, K. Z. (1981) *The foundations of ethology*. New York: Springer-Verlag.
15. Lorenz: *The foundations of ethology. op. cit*.; and Livesey, P. J. (1986) *Learning and emotion: a biological synthesis*. London: Lawrence Erlbaum Associates.
16. See Popper, K. R. (1972) *Objective knowledge—an evolutionary approach*. Oxford: Clarendon; Dennett, D. C. (1995) *Darwin's Dangerous Idea*. New York: Simon and Schuster; and Stewart: Metaevolution. *op. cit*.
17. See Boyd, R. and P. J. Richerson (1985) *Culture and the evolutionary process*. Chicago: University of Chicago Press.
18. Stewart: Metaevolution. *op. cit*.
19. Laland, K. N. (1992) A Theoretical Investigation of the Role of Social Transmission in Evolution. *Ethology and Sociobiology* **13**: 87-113.

Chapter 11 Smarter Humans

1. Stewart, J. E. (1997) Evolutionary Progress. *Journal of Social and Evolutionary Systems* **20**: 335-362.
2. Stewart, J. E. (1995). Metaevolution. *Journal of Social and Evolutionary Systems* **18**: 113-147; and Stewart: Evolutionary Progress. *op. cit*.
3. Stewart: Metaevolution. *op. cit*; and Stewart: Evolutionary Progress. *op. cit*.

Chapter 12 The Self-Evolving Organism

1. For example, see Vernon, P. E. Ed. (1970) *Creativity.* Harmondsworth: Penguin.
2. Stewart, J. E. (1997) Evolutionary Progress. *Journal of Social and Evolutionary Systems* **20**: 335-362.
3. For example, see Wilson, C. (1956) *The outsider.* London: Gollancz.
4. Stewart, J. E. (1995) Metaevolution. *Journal of Social and Evolutionary Systems* **18**: 113-147; and Stewart: Evolutionary Progress. *op. cit.*
5. The clearest and most practical description of Gurdjieff's techniques and practices for achieving psychological transformation are Nicoll, M. (1980) *Psychological Commentaries on the Teaching of Gurdjieff and Ouspensky.* Volumes 1 to 5. London: Watkins.

Chapter 13 The Evolution of Life on Earth

1. See Kauffman, S. A. (1993) *The Origins of Order: Self-organisation and selection in evolution.* New York: Oxford University Press; and Kauffman, S. A. (1995) *At home in the universe: The search for laws of self-organisation and complexity.* New York: Oxford University Press.
2. Cavalier-Smith, T. (1981) The origin and early evolution of the early eukaryote cell. In: *Molecular and Cellular Aspects of Microbial Evolution. Society for General Microbiology Symposium 32.* (Carlisle, M. J. *et al.*, eds.) pp. 33-84. Cambridge: Cambridge University Press.
3. Stewart, J. E. (1997) Evolutionary Progress. *Journal of Social and Evolutionary Systems* **20**: 335-362.
4. But insect societies have evolved some processes that adapt the society as a whole. For example, see Bonabeau, G. T., Deneubourg, J., Aron, S. and S. Camazine. (1997) Self-organization in social insects. *Trends in Ecology and Evolution* **12**: 188-193.
5. Holldobler, B. and E. O. Wilson (1990) *The Ants.* Cambridge, MA:

Harvard University Press.

Chapter 14 Management by Morals

1. For a discussion of moral systems from an evolutionary perspective, see Alexander, R. D. (1987) *The biology of moral systems.* New York: Aldine de Gruyter.
2. Stewart, J. E. (1997) Evolutionary transitions and artificial life. *Artificial Life* **3**: 101-120.
3. See Klein, R. G. (1989) *The Human Career: Human Biological and Cultural Origins.* Chicago: University of Chicago Press.
4. Boehm, C. (1997) Impact of the human egalitarian syndrome on Darwinian selection mechanics. *The American Naturalist.* **150**: S100-S121.
5. See, for example, Rodseth, L., Wrangham, R. W., Harrigan, A. M. and B. B. Smuts. (1991) The human community as a primate society. *Current Anthropology* **32**: 221-254.
6. Trivers, R. (1985) *Social Evolution.* Menlo Park, CA: Benjamin/Cummings.
7. This operates on exactly the same basis as the kin selection mechanism. In kin selection a gene organises an individual to benefit other individuals that also contain a copy of the gene. Here an inculcated belief does the same.
8. Alexander: *The biology of moral systems. op. cit.*
9. See Chapter 4 of Ridley, M. (1996) *The Origins of Virtue.* London: Viking.
10. Rappaport, R. A. (1979) *Ecology, meaning and religion.* Richmond, CA: North Atlantic Books.
11. *Ibid.*
12. For examples see Colson, E. (1974) *Tradition and contract: the problem of order.* Chicago: Adeline Publishing; and Posner, R. (1980) A theory of primitive society, with special reference to the law. *Journal of Law and Economics.* **XXIII**: 1-54.
13. Numerous detailed examples of how religious systems have mu-

tated in this way are given by La Barre, W. (1970) *The Ghost Dance: Origins of Religion.* London: George Allen and Unwin.

Chapter 15 The Rise of Governed Societies

1. See, for example, Johnson, A. W. (1987) *The evolution of human societies: from foraging group to agrarian state.* Stanford, California: Stanford University Press.
2. Stewart, J. E. (1995) Metaevolution. *Journal of Social and Evolutionary Systems* **18**: 113-147.
3. Stewart, J. E. (1997) Evolutionary Progress. *Journal of Social and Evolutionary Systems* **20**: 335-362.
4. The ability of cultural norms and religious beliefs to facilitate co-operative economic organisation is discussed in detail by North, D. C. (1990) *Institutions, institutional change and economic performance.* New York: Cambridge University Press.
5. It is worth noting here that what I refer to as the management of a society includes most of what economists refer to as 'institutions'. They are the controls that constrain how members of the society can pursue their interests.
6. The evolution and emergence of market systems is dealt with by North, D. C. (1991) Institutions. *Journal of Economic Perspectives* **5**: 97-112.
7. See, for example, Eidem, R. and S. Viotti (1978) *Economic Systems.* Oxford: Martin Robertson and Co.
8. See Hayek, F. A. (1948) *Individualism and Economic Order.* Chicago: University of Chicago Press.
9. The way in which consumers provide the selective environment for the activities of producers is set out particularly clearly in the work of evolutionary economists. See, for example, Nelson, R. and S. Winter (1982) *An evolutionary theory of economic change.* Cambridge: Cambridge University Press.
10. For a fuller discussion of the limitations of bureaucratic organisation see March, J. G. and H. A. Simon. (1971) The dys-

functions of bureaucracy. In *Organisation theory,* (Pugh, D. S. ed.) pp 30-42. Harmondsworth: Penguin Education.

11. See, for example, Ritschel, D. (1997) *The Politics of Planning: The Debate on Economic Planning in Britain in the 1930s.* (Oxford Historical Monographs) Oxford: Clarendon Press.

Chapter 16 Limitations of Markets

1. For a comprehensive discussion of what economists call public goods see Buchanan, J. and G. Tulloch (1962) *The Calculus of Consent.* Ann Arbour: University of Michigan Press.
2. Brittan, S. (1983) *The role and limits of government.* Minneapolis: University of Minnesota Press.
3. See, for example, Mills, E. S. (1978) *The economics of environmental quality.* New York: W. W. Norton.
4. Thurow, L. (1997) *The Future of Capitalism.* p243. St Leonards: Allen and Unwin.
5. See, for example, Taylor, A. J. (1972) *Laissez-faire and State Intervention in Nineteenth Century Britain.* London: Macmillan.
6. For a fuller discussion see Marcuse, H. (1964) *One dimensional man: studies in the ideology of advanced industrial society.* London: Routledge & Keegan Paul.
7. This argument is made out in detail in Bowles, S. and H. Gintis (1993) The revenge of homo economicus: contested exchange and the revival of political economy. *Journal of Economic Perspectives.* **7**: 83-102.
8. For a more detailed discussion of these issues, see Schumpeter, J. A. (1947) *Capitalism, Socialism, and Democracy.* New York: Harper & Bros.
9. See, for example, Bork, R. H. (1978) *The Antitrust Paradox.* New York: Basic books.
10. Thurow: *The Future of Capitalism. op. cit.*
11. A very good analysis of the ways in which ideology can control society is found in Chapter 5 of Mc Murtry, J. (1978) *The Struc-*

ture of Marx's World-View. Princeton, New Jersey: Princeton University Press.

12. Again, a very good analysis of the ways in which the economically powerful can influence ideas and beliefs is to be found in Mc Murtry: *The Structure of Marx's World-View. op. cit.*
13. *Ibid.*

Chapter 17 Planetary Society and Beyond

1. The globalisation of markets is discussed in depth in Carnoy, M. (1993) *The new global economy in the information age.* University Park: Pennsylvania State University Press.
2. See, for example, Held, D. (1995) *Democracy and the global order.* Cambridge: Polity Press.
3. The need for global governance to deal effectively with global environmental problems is discussed at length by Lipschutz, R. D. (1996) *Global civil society and global environmental governance: the politics of nature from place to planet.* Albany: State University of New York Press.
4. Several advantages of global governance including its ability to prevent war are discussed in the final Chapter of Jones, W. S. (1991) *The logic of international relations.* New York: HarperCollins Publishers.
5. For example, the necessity of global governance to prevent human rights abuse within countries is demonstrated in detail by Robertson, G. (1999) *Crimes against humanity: the struggle for global justice.* London: Allen and Lane.
6. This case is argued strongly in Ohmae, K. (1995) *The end of the nation state: the rise of regional economies.* London: HarperCollins.
7. The effects of globalisation on wage levels in developed countries are discussed in Freeman, R. (1995) Are your wages set in Peking? *Journal of Economic Perspectives* **9**: Summer.
8. See, for example, Gray, J. (1998) *False Dawn: the Delusions of*

Global Capitalism. London: Granta Books.

9. See, for example, Mander, J. and E. Goldsmith (1996) *The case against the global economy and for a turn toward the local.* San Francisco: Sierra Books.
10. The European Union is discussed in the context of moves to global governance by Giddens, A. (1998) *The third way.* Cambridge: Polity Press.
11. The idea of a vertical market was first developed in detail in Stewart, J. E. (1995) Metaevolution. *Journal of Social and Evolutionary Systems* **18**: 113-147; and Stewart, J. E. (1997). Evolutionary Progress. *Journal of Social and Evolutionary Systems* **20**: 335-362.
12. Markets in permits to produce carbon dioxide emissions are discussed in Ingham, A. and A. Ulph (1991) The economics of global warming. In: *Reconciling economics and the environment.* (Bennet, J. and Block, W. eds.) pp. 223-248. Perth: The Australian Institute for Public Policy.
13. For more on ideas futures markets see Hanson, R. (1995) Ideas Futures. *Wired.* September: p125.

Chapter 18 Humanity V. Bacteria

1. Ayala, F. J. (1997) Ascent by natural selection. *Science* **275**: 495-6.
2. Gould, S. J. (1994) The Evolution of Life on Earth. *Scientific American* October: 63- 69.
3. A number of examples of cooperation between bacteria and other organisms are examined in Margulis, L. and D. Sagan (1986) *Microcosmos.* New York: Summit.
4. See, for example, Clapham, W. B. (1973) *Natural Ecosystems.* New York: Macmillan.
5. Stewart, J. E. (1997) Evolutionary Progress. *Journal of Social and Evolutionary Systems* **20**: 335-362.
6. *Ibid.*

7. See Drexler, K. E. (1986) *Engines of creation: the coming era of nanotechnology.* New York: Doubleday Anchor.

8. Adams, F. and Laughlin, G. (1999) *The Five Ages of the Universe.* New York: Free Press.

9. Stewart: Evolutionary Progress. *op. cit.*

Index

A

B

C

D

E

F

G

R

S

T

U

V

W

Y

CPSIA information can be obtained at www.ICGtesting.com
Printed in the USA
BVOW081127181112

305848BV00001B/134/A